Hydropower: An Introduction

Hydropower: An Introduction

Stephen Adams

MURPHY & MOORE
www.murphy-moorepublishing.com

Murphy & Moore Publishing,
1 Rockefeller Plaza,
New York City, NY 10020, USA

Visit us on the World Wide Web at:
www.murphy-moorepublishing.com

ISBN: 978-1-63987-319-7 (Hardback)

Cataloging-in-Publication Data

Hydropower : an introduction / Stephen Adams.
p. cm.
Includes bibliographical references and index.
ISBN 978-1-63987-319-7
1. Water-power. 2. Power resources. 3. Hydroelectric power plants.
I. Adams, Stephen.
TC147 .H93 2022
621.312 134--dc23

Table of Contents

Preface

Hydropower is a form of renewable energy, which is derived from falling or fast-running water. The electricity produced by hydropower is known as hydroelectricity. Hydroelectric power plants are used to exploit the energy of falling water and generate hydroelectricity. They can also be used to store potential energy between two reservoirs at different heights. Hydropower finds extensive application in irrigation and operations conducted in ore mills, textile mills, gristmills and sawmills. This book attempts to understand the multiple branches that fall under the discipline of hydropower and how such concepts have practical applications. Most of the topics introduced in this book cover new techniques and the applications of hydropower. Coherent flow of topics, student-friendly language and extensive use of examples make this book an invaluable source of knowledge.

A foreword of all chapters of the book is provided below:

Chapter 1 - The power derived from falling or fast-running water that may be operated for useful purposes is known as hydropower. A hydropower system uses a dam to store river water in a reservoir. Water released from the reservoir flows through a turbine which spins it activating generator to produce electricity. This is an introductory chapter which will introduce briefly all the significant aspects of hydropower; **Chapter 2 -** A dam is a large, man-made barrier which restricts the flow of surface water or underground streams. On the basis of structure, dams are of seven types: arch dams, gravity dams, arch-gravity dams, barrages, embankment dams, rock-fill embankment dams and fixed-crest dams. Dams have various other types based on their size, use and material. The topics elaborated in this chapter will help in gaining a better perspective about the types of hydroelectric dams; **Chapter 3 -** The electricity produced by generators that are pushed by the movement of water is known as hydroelectricity. The method of using mechanical energy of water through man-made conduits to generate electricity is known as conduit hydroelectricity. This chapter has been carefully written to provide an easy understanding of the varied facets of hydroelectricity; **Chapter 4 -** Watermill is a mill that uses a water turbine to operate a mechanical process. A water turbine is a rotary machine that converts the kinetic and potential energy of water into mechanical or electrical energy. All the diverse aspects of watermills and water turbines have been carefully analyzed in this chapter; **Chapter 5 -** Ocean energy encompasses all forms of renewable energy carried by ocean waves, salinity and tides. The energy which is generated by using the kinetic energy from the rise and fall of tides is known as tidal energy. This chapter closely examines the key concepts of ocean energy to provide an extensive understanding of the subject.

I would like to thank the entire editorial team who made sincere efforts for this book and my family who supported me in my efforts of working on this book. I take this opportunity to thank all those who have been a guiding force throughout my life.

Stephen Adams

Chapter 1

Hydropower as an Energy Source

The power derived from falling or fast-running water that may be operated for useful purposes is known as hydropower. A hydropower system uses a dam to store river water in a reservoir. Water released from the reservoir flows through a turbine which spins it activating generator to produce electricity. This is an introductory chapter which will introduce briefly all the significant aspects of hydropower.

Hydropower is a renewable energy source where power is derived from the energy of water moving from higher to lower elevations. So hydropower is the energy contained in water (i.e. electricity from water). Hydropower is a renewable energy source where power is derived from the energy of water using its gravitational force when it is flowing or falling. It is a proven, mature, predictable and cost competitive technology. Hydropower has among the best conversion efficiencies of all known energy sources (about 90% efficiency, water to wire). It requires relatively high initial investment, but has a long lifespan with very low operation and maintenance costs. In the past, hydropower has acted as a catalyst for economic and social development by providing both energy and water management services, and it can continue to do so in the future. Hydro storage capacity can mitigate freshwater scarcity by providing security during lean flows and drought for drinking water supply, irrigation, flood control and navigation services. Hydropower plants do not consume the water that drives the turbines. The water, after power generation, is available for various other essential uses. Hydropower can serve both in large, centralized and small, isolated grids, and small-scale hydropower is an option for rural electrification.

Hydropower Potential

Hydropower potential may be estimated by using equation:

$$\text{Power in kW (P)} = Q.H.9.81\,\eta$$

Where,

- Q: Discharge in m^3/s,
- H: Head in meters,
- η: Overall efficiency of turbine, generator and gear box (may be taken as 75% initially).

The head is relatively constant in run-of-river schemes except for variation in friction losses with the varying discharge. In irrigation canal or dam toe based scheme head also vary depending season of release of water as water level in the dam varies throughout the year. Weighted average is found out to design the head of turbine. The design head is so selected that turbine is operated to the maximum time giving optimum energy generation. Energy per year is calculated as below:

E = P.time (in hours 8760 per year) in kiloWatt hour

For the run of river hydropower scheme, it is useful to know the variation of flow over the year to select the most appropriate turbine configuration and estimated power generation. Flow variation presented in the form of a flow duration curve is the most useful form. The firm power and secondary power could be easily estimated from the flow duration curve.

Power potential may not be necessary developed fully as it all depends upon the economics of power project, power demand, grid characteristics. If the small hydro station is connected to the grid, the entire energy generated can be used by the system. For isolated operation load forecast is necessary.

The potential for energy production in a hydropower plant is determined by the following parameters, which are dependent on the hydrology, topography and design of the power plant:

- The amount of water available.
- Water loss due to flood spill, bypass requirements or leakage.
- The difference in head between upstream intake and downstream outlet.
- Hydraulic losses in water transport due to friction and velocity change.
- The efficiency in energy conversion of electromechanical equipment.

In addition, some energy losses occur in the water conductor system from diversion point (intake) to turbine. These losses, called head loss, reduce the head and hence the energy potential for the power plant. These losses can be classified either as friction losses or singular losses. Friction losses depend mainly on water velocity and the roughness in tunnels, pipelines and penstocks.

The total efficiency of a hydropower plant is determined by the sum of these three loss components. Hydraulic losses can be reduced by increasing the turbine capacity or by increasing the reservoir capacity to get better regulation of the flow. Head losses can be reduced by increasing the area of headrace and tailrace, by decreasing the roughness in these and by avoiding too many changes in flow velocity and direction. Different turbine types have quite different efficiency profiles when the turbine discharge deviates from the optimal value.

Technology

Head and also installed capacity (size) are often presented as criteria for the classification of hydropower plants. A broad range of hydropower systems, classified by project type, system, head or purpose, can be designed to suit particular needs and site-specific conditions. Hydro power is broadly classified in Micro, Mini, Small, Medium and Large Hydro systems based on plan capacity. Micro is up to and below 100 kW, mini is up to 2 MW, small is up to 25 MW, medium is up to 100 MW and large is above 100 MW. Further hydropower also classified based on head available as Ultra Low Head is below 3 metres, Low Head is less than 40 metres, Medium Head is above 40 metres and High head is above 75 metres.

Hydropower plants are often classified in three main categories according to operation and type of flow. Run-of-river (RoR), storage (reservoir) and pumped storage HPPs all vary from the very small to the very large scale, depending on the hydrology and topography of the watershed. In addition, there is a fourth category called in-stream technology, which is a young and less-developed technology.

Run of River Scheme

These are hydropower schemes in which the available water is diverted from stream for hydro power generation. Such a hydropower plant generally includes some short-term storage (hourly, daily, or weekly), allowing for some adaptations to the demand profile. Run-of-river hydropower plants are normally operated as base-load power plants. A portion of river water might be diverted to a channel, pipe line (penstock) to convey the water to hydraulic turbine which is connected to an electricity generator. Installation of small RoR plants is relatively cheap and has in general only minor environmental impacts.

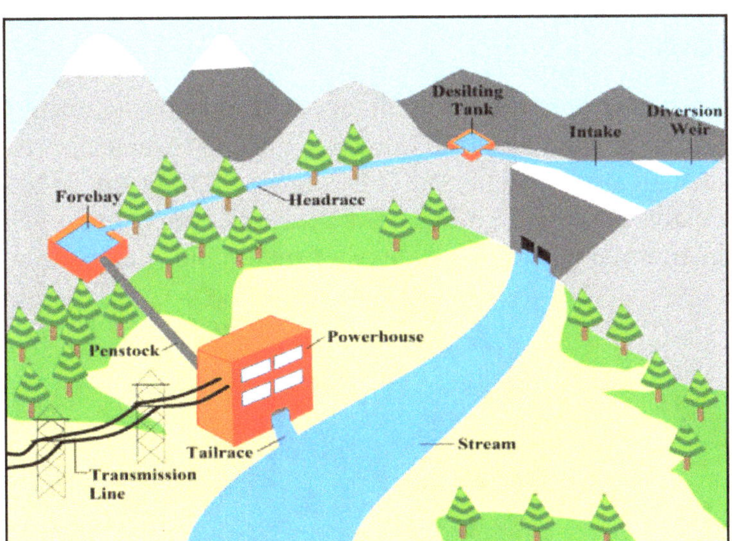

Run of river hydropower plant.

Reservoir

In order to reduce the dependence on the variability of inflow, many hydropower plants comprise reservoirs where the generating stations are located at the dam toe or further downstream through tunnel or pipelines as per the electricity or downstream water demand. Such reservoirs are often situated in river valleys. Most of the time these dams are constructed in the river for desired use like irrigation, drinking, flood control.

Hydropower plants with reservoir.

Pumped-storage

Pumped-storage plants pump water into an upper storage basin during off-peak hours using surplus electricity from baseload power plants and reverse flow to generate electricity during the daily peak load period. It is considered to be one of the most efficient technologies available for energy storage. The excess electricity in the grid is usually the generation of the thermal power plants or other renewable energy e.g. wind and solar.

Pumped storage project.

Instream Technology using Existing Facilities

To optimise existing facilities like weirs, barrages, canals or falls, small turbines can be installed for electricity generation. Smaller falls are also available on cooling water return channels on thermal power stations, drinking water and sewage channel after treatment. These are basically functioning like a run-of-river scheme.

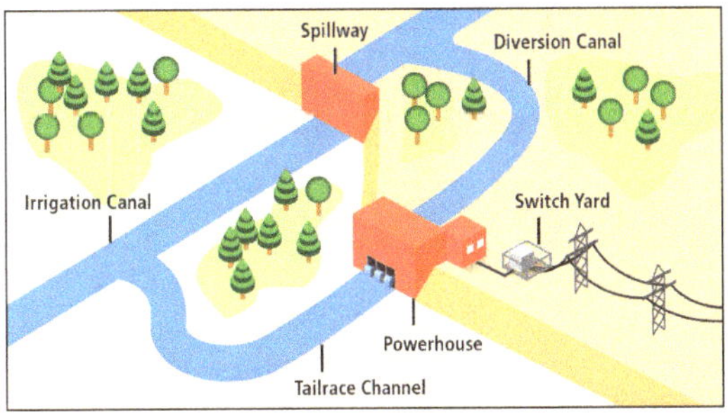

Typical arrangement of instream technology hydropower projects.

Advantages of Hydro Power

- Clean fuel: Hydropower is fueled by water.
- The SHPs (Small/mini/micro hydel projects) projects have potential to meet power requirements of remote and isolated areas - most attractive renewable source of grid quality power generation.
- Conserve fossil fuels.
- Substitute thermal power thereby reducing carbon emissions.
- Hydropower relies on the water cycle, which is driven by the sun, thus it's a renewable power source.
- Hydropower is generally available as needed; Engineers can control the flow of water through the turbines to produce electricity on demand.
- Hydropower plants provide benefits in addition to clean electricity.
    - Water supply,
    - Flood control,
    - Impoundment hydropower creates reservoirs.

That offers a variety of recreational activities: Fishing, swimming, and boating. It requires relatively high initial investment, but has a long lifespan with very low operation and maintenance costs.

Disadvantages of Hydro Power

The local social and environmental impacts of hydropower projects vary depending on the project's type, size and local conditions. Some impacts such as changes in flow and water quality of river, barriers to fish migration, impact on biological diversity, displacement of people due to impoundments. But reservoirs provide, beyond electricity supply, multiple beneficial services. While lifecycle assessments indicate very low carbon emissions, there is currently no consensus on the issue of land use change-related net emissions from reservoirs. In the past, hydropower has acted as a catalyst for economic and social development by providing both energy and water management services, and it can continue to do so in the future. Hydropower can serve both in large, centralized and small, isolated grids, and small-scale hydropower is an option for rural electrification.

Hydropower Plants

Hydroelectric Power Plant is a system in dams and works by obstructing the river flow which causes to increase and store water in the Dams. This is Potential Energy. The water is made to fall from a height which constitutes Kinetic Energy. This Kinetic Energy is then converted to Mechanical Energy by the Turbines. Generator is responsible in converting this Mechanical Energy from the Turbine into Electrical Energy. Hydroelectric Power Plants generates electricity for home consumption or business needs. The flexibility of generating the electricity usually is either in a large scale or in a smaller scale depending on the usage. In 1879, the first commercial Hydroelectric Power Plant was built at Niagara Falls. Although every passing year, there has been continuous improvements doing rounds for safer and efficient utilization.

Types of Hydropower Plants

There are different types of hydro power plants based on types of facilities for generation of hydropower. Construction of large hydropower plants is practical and economically viable proposition as the capital costs of a project can be reduced with such installations.

With the growth of power systems, there has been increase in the need for peak load capability of hydropower plants. In some power plant systems, the ratio of power demands at peak periods and off-peak periods has also become greater. The use of reversible pump turbines is one of the method by which the growing demands for economical peak power is being met. The evolution of the bulb turbines or pump turbines has helped in greater economy in the development of low-head River and tidal power projects.

Different Types of Modern Hydro Power Plants

Pumped Storage Hydropower Plants

To supply the peak loads, hydropower plants has to have the installed capacity of high loads of which remains idle during the off-peak hours. The more the demands of variable power supply, it is necessary to devise some way to achieve the economical loading of the power plant by leveling up the load curve. The following are some of the ways:

- Commercial Method: To sell electric current at a higher rate during peak hours than during off-peak hours.
- Technical Method: The following are the two methods:
    - By installing special peak load power plants.
    - By storing energy produced during off-peak hours. Such a system is known as Pumped Storage Plants.

Purpose of Pumped Storage Hydropower Plants

This type of plants combined with steam power stations reduces the power load fluctuations to narrow limits. In some cases, the storage plant consists of pump and motor with no turbines. The pump increases the head in the feeder reservoir of a separate hydro-electric plant while motor improves the power factor in the electric supply network. Combined with hydro-electric power plants, it can take charge of daily peak load and seasonal variations of water.

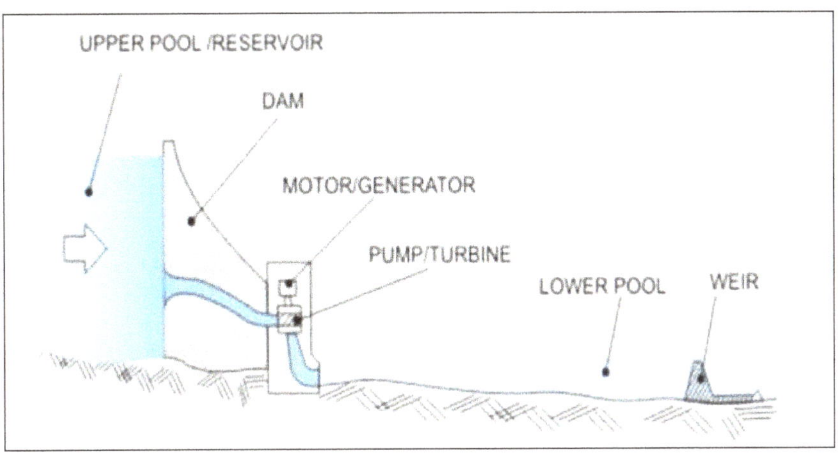

Types of Pumped Storage Plants:

- Daily, weekly or seasonal storage plants.
- High, medium or low head plants.
- According to the type of turbine used in the plant.
- Pure or mixed storage hydropower plants.
- Horizontal or vertical storage plants.

Reversible Turbine Hydro Power Plants

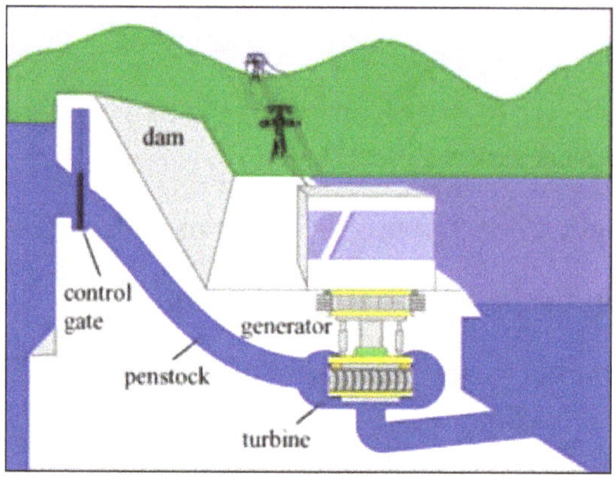

In this type of hydropower plants, Francis turbines which are just a reverse of centrifugal pump functions in one direction as a motor driven pump and in the reverse direction as turbine. The efficiencies of pumping and generating processes have been found to exceed 65%. As the specific speed of a pump is greater than that of a geometrically similar turbines, two different rotational speeds are necessary for the two machines, if the best efficiency is to be obtained in each direction of rotation. The same speed is possibly only with some sacrifice of efficiency.

Underground Hydropower Plants

The underground hydropower plants are constructed below the upper level of water. These are low head plants. This type of installation consists of a weir across the river to be harnessed and housing the power plants within the weir.

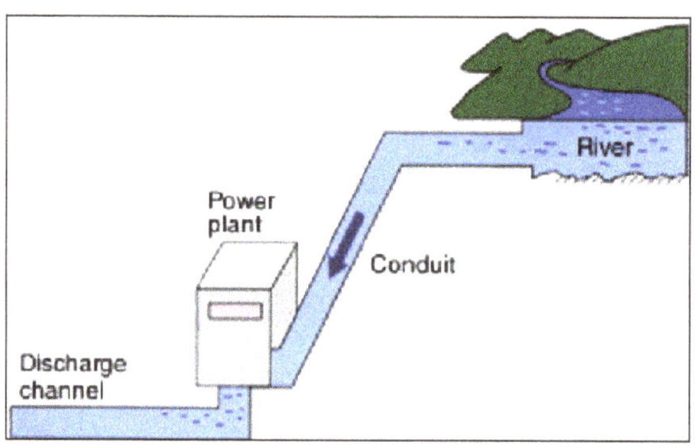

The head race is connected to the tail race by a straight passage in which the tubular or bulb turbines operate. The turbine is connected to the generator by a horizontal or an inclined shaft depending on convenience. The advantages of underground hydropower plants are:

- They are safe against any attacks by enemies or terrorists, rock and earth slips and snow avalanches.

- Air-conditioning is possible throughout the year.
- They show greater operating heads.
- Since excavation is in rocks, considerable savings in the cost of power plant construction and erection of accessories are possible.
- They do not affect the scenic beauty of the site.

Tidal Power Plants

Tidal power plants are not considered as a hydropower plant. The discharge and head are made available due to the tide. This is due to rotation of earth.

Tides flood vast areas and raise the sea level along the seashores by amounts varying from a few centimeters to several meters. A tremendous amount of energy is stored in the earth in this process. The force exerted by tides is such as to slow down the earth's rotation.

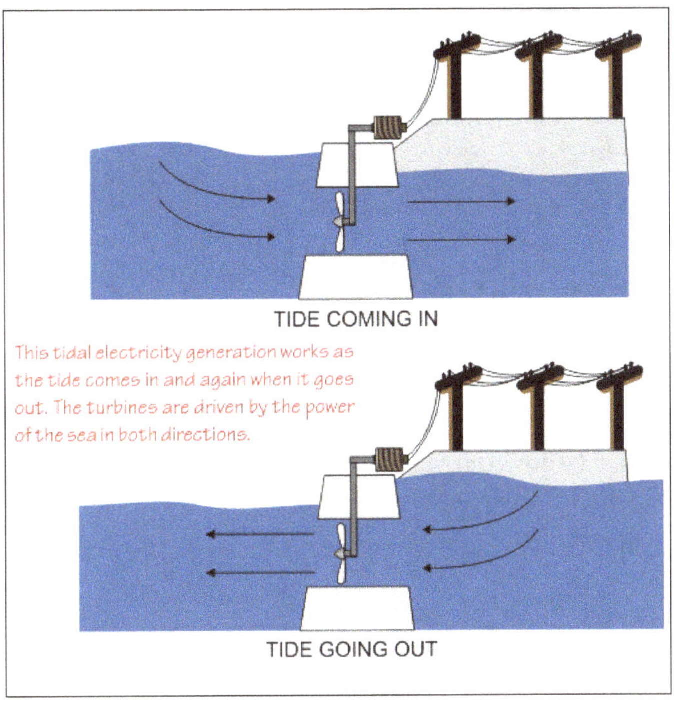

Tidal energy can be increased by artificial arrangements, such as:

- If the gates connecting the tidal basin with the sea are suddenly opened at high water, the tidal wave rapidly begins to raise the basin level. After reflection, the filling wave returns to the dam, its crest being above the initial sea level. If the gates can be quickly closed exactly at this instant, the plant can operate during the ebb under an additional head created by reflection of this flood wave.

- Pumping into the basin from the sea at high water and from the basin into the sea at low water is also another method of proper utilization of tidal power.

Advantages of Tidal Power Plants

- To compensate the failure of hydro-power plants.
- To provide constant power supply.
- To improve the efficiency of entire power generating system.
- To operate the atomic and thermal power plants under most economical conditions.
- These are cheaper than thermal or hydropower plants.

Components of Hydropower Plant

A hydro power plant comprises of following components:

- Civil components.
- Hydro electrical components comprising of hydro mechanical and hydro electrical equipment.

Civil Components

Diversion Structure or Reservoir

The function of diversion structure is to divert the water required for power generation to the sedimentation tank through feeder channel whereas the function of reservoir is to store the water from surplus period to deficit period. Reservoir or diversion structure may be built on surface (weir &barrage, inflatable rubber dam). The structure designed should be safe against sliding, overturning and piping.

Desilting Device

For run of river type of schemes, desilting devices are provided in the water conductor system as diverted water carries sediments as may damage the turbine parts, penstock and other components due to abrasion. In the desilting devices the incoming velocity of water is lowered by increasing the cross sectional area of desilting device thereby effectively allowing the sediment of particle.

Power Channel

The power channel is the one which carries water from source to forebay tank. The main function of power channel is for water conveyance, and sometimes to take the surge of water. The power channel may be lined with concrete to prevent the loss of water due to seepage. The shape of the power channel may be of rectangular or trapezoidal.

Forebay Tank or Surge Tank

The forebay is a small storage pond located in between power channel and penstocks. It provides immediate water demand on starting the generating units and required water seal over the penstock inlet against air entrainment. Surge tank is the storage with possibility of allowing the surge to more up and down without causing any penstock. Suitable spillway is provided on one side of forebay to dispose of safely the excess inflows during load rejection.

Penstock

Pressure conduits carrying water from forebay/surge tank to the power house are termed as penstock. Penstocks are required to bear maximum water pressure including water hammer which occurs due to sudden closure of inlet valve; they are quite costly and are a very important part of the water conductor system. The penstock material can be of steel, glass fiber reinforced plastic or concrete.

Power House

The functions of the power house are to support and house the generating units (turbine and its auxiliary units) and their accessories against weather and other human activities. It can be of surface type or underground type also.

Tail Race Channel

After passing through the turbines, the water is discharged back into the stream through a short channel called tailrace. The tailrace channel to be designed ion such a way that minimum tail water level is required to maintain safe suction head for smooth operation of turbine.

Hydro Mechanical Equipment

Hydro mechanical equipment comprises of Hydro turbine, gear box and gates & valves.

Hydro Turbine

Hydro Turbine is a fluid machine used for converting hydro potential available in water into mechanical power and then utilizes this power for driving the electric generator in power plant. Turbines can be either reaction or impulse types. The turbines type indicates the manner in which the water causes the turbine runner to rotate. Reaction turbine operates with their runners fully flooded and develops torque because of the reaction of water pressure against runner blades. Reaction turbines are classified as

Francis (mixed flow) or axial flow. Axial flow turbines are available with both fixed blades (Propeller) and variable pitch blades (Kaplan). Both axial flow (Propeller & Kaplan) and Francis turbines may be mounted either horizontally or vertically. Additionally, Propeller turbines may be slant mounted.

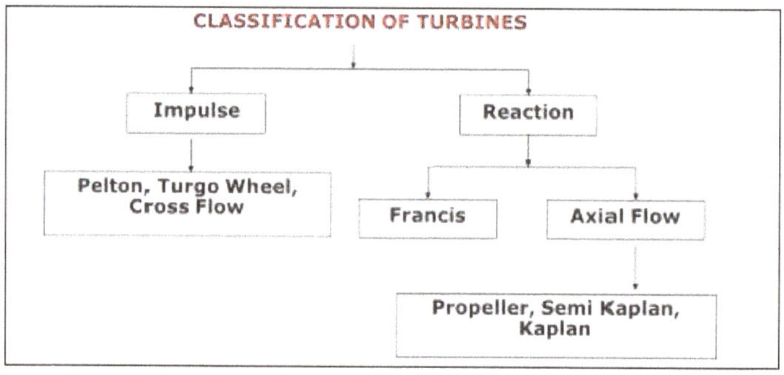

Classification of Hydro Turbines.

Second type of turbines are Impulse turbines and operate with their runner in air and convert the water's pressure energy into kinetic energy of a jet that impinges onto the runner buckets to develop torque. Pelton turbines are impulse turbines and installed at high head. The efficiency of different hydraulic turbines may be seen in figure.

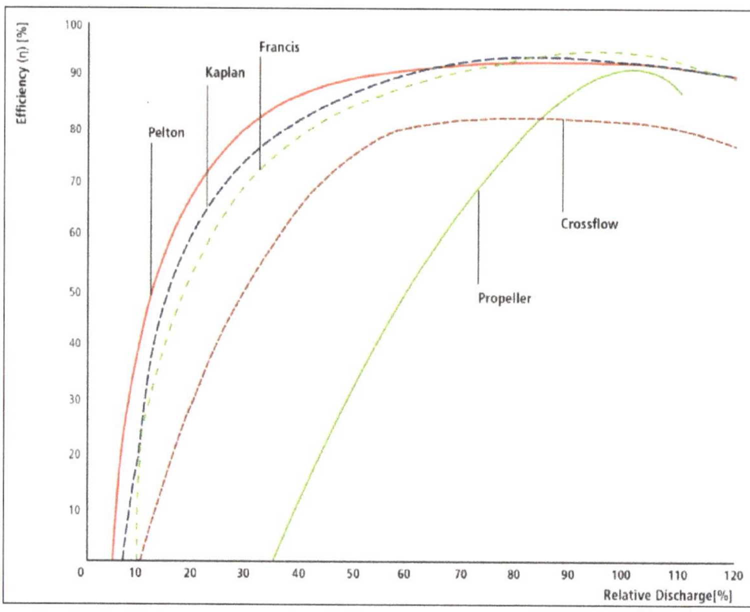

Typical efficiency curves for different types of hydropower turbines.

Speed Governors

A speed governor is a combination of devices and mechanisms, which detect speed deviation and convert it into a change in servomotor position. A speed-sensing element detects the deviation from the set point; this deviation signal is converted and amplified to excite an actuator, hydraulic or electric, that controls the water flow to the turbine. In a Francis turbine, where there is a reduction in water flow needed to rotate the wicket-gates. For this, a powerful governor is required to overcome the hydraulic and frictional forces and to maintain the wicket-gates in a partially closed position or to close them completely. Several types of governors arc available, varying from old fashioned purely mechanical to mechanical-hydraulic to electrical-hydraulic and mechanical-electrical.

Speed Increasers

When the turbine and the generator operate at the same speed and can be placed so that their shafts are in line, direct coupling is the right solution; virtually no power losses are incurred and maintenance is minimal. Turbine manufactures recommend the type of coupling to be used, either rigid or flexible although a flexible, coupling that can tolerate certain misalignment, is usually recommended.

In many instances, particularly in low head schemes, turbines run at less than 400 rpm, requiring a speed increaser to meet the 750-1000 rpm of standard alternators. In the range of powers contemplated in small hydro schemes, this solution is often more economical than the use of a custom made alternator.

Hydro Electrical Equipment

It comprises of Electrical generators, power and instrument Transformers, circuit breakers and relays, etc.

Electrical Generator

It converts the mechanical energy available from hydro turbine into electrical energy by means of electromechanical energy conversion process. Turbine speed governs the design of generator. The synchronous speed of generator is given by:

$N_S = 120\ f/P$

Where, P is number of poles and f is the frequency of system.

Mostly two types of generators are used in electrical power systems which are:

- Synchronous generators: They are equipped with a DC electric or permanent magnet excitation system (rotating or static) associated with a voltage regulator to control the output voltage before the generator is connected to the grid. They supply the reactive energy required by the power system when the generator is connected to the grid. Synchronous generators can run isolated from the grid and produce power since excitation is not grid-dependent.

- Asynchronous generators: They are simple squirrel-cage induction motors with no possibility of voltage regulation and running at a speed directly related to system frequency. They draw their excitation current from the grid, absorbing reactive energy by their own magnetism. Adding a bank of capacitors can compensate for the absorbed reactive energy. They cannot generate when disconnected from the grid because are incapable of providing their own excitation current. However, they are used in very small stand-alone applications as a cheap solution when the required quality of the electricity supply is not very high.

In synchronous generator speed (Ns) is kept constant whereas in Asynchronous generator speed (Ns) is varying.

Transformers

It is a static device used to transfer electrical energy from its primary to secondary side by stepping up or steeping down the voltage. Step up transformers are used at power stations to step up the generated voltage for the purpose of transmission, transmitting higher voltage over long distance has many advantages like low power transmission loss, reduction in cable and insulator size, low cost and further. Step down transformer are used at distribution end to step down the voltage according to requirement of users.

Instrument Transformer

It is a measuring instrument used to measure the high voltage and high current in electrical power systems. These are categorised into two types which are:

- Current transformers,
- Voltage transformers.

Circuit Breakers

It is a protective device which protects the electrical networks from faulty condition. The main purpose of circuit breaker is to isolate the faulty parts from the networks so that fault is being localised. Earlier fuse was used now it's being replaced by circuit breakers.

Relay

It is also a protective device which used to sense the abnormal condition in the electrical system. Whenever abnormal condition occurs relay operates by closing its contacts to give the signal to circuit breaker for opening circuit breaker contacts there by fault is isolated.

Powerhouse

In a hydro power scheme the role of the powerhouse is to protect the electromechanical equipment that converts the potential energy of water into electricity from the weather and provide a place for carrying out operation and maintenance activities. The number, type and power of the generating units, their configuration, the scheme head and the topographical conditions of the site determine the shape and size of the building. The layout may differ from project to project as per site conditions. The powerhouse shall comprise of machine hall having main hydro generating equipment (turbine, governor and generator), service bay to carrying out erection and maintenance activities, control room having control panels, relays etc. and tailrace for water exit.

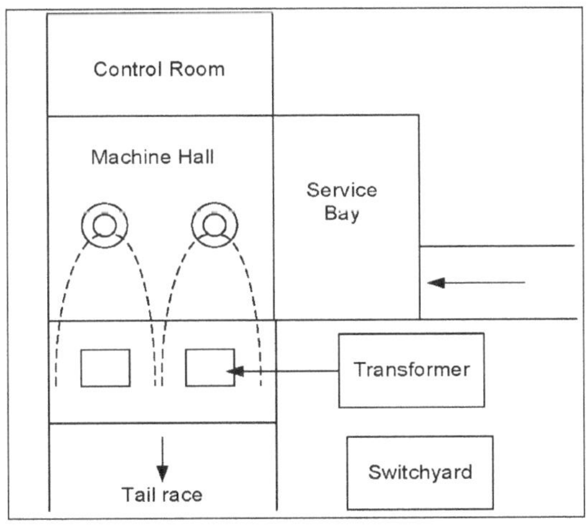

Following equipment are placed in the powerhouse building:

- Inlet gate or valve,
- Turbine,
- Speed increaser (if needed),
- Generator,
- Control system,
- Condenser, switchgear,
- Protection systems,
- DC emergency supply,
- Power and current transformers etc.

References

- Hydropower-as-an-energy-source, Energy-and-Environment: inflibnet.ac.in, Retrieved 04, January 2020
- Types-of-hydro-power-plants, water-resources -9554: theconstructor.org, Retrieved 27, April 2020

Chapter 2

Hydroelectric Dams

A dam is a large, man-made barrier which restricts the flow of surface water or underground streams. On the basis of structure, dams are of seven types: Arch dams, Gravity dams, Arch-gravity dams, Barrages, Embankment dams, Rock-fill embankment dams and Fixed-crest dams. Dams have various other types based on their size, use and material. The topics elaborated in this chapter will help in gaining a better perspective about the types of hydroelectric dams.

A dam is any retaining structure, built across a river to stop or regulate its flow, and to raise the water level. Once a dam is built, water accumulates behind it to form a lake or dam reservoir. Thus it also includes containment structures for a large or small body of impounded water.

While dams are built to stop flooding and facilitate shipping, they may also be intended to generate electricity, and these are called hydroelectric dams. Such a structure is an essential part of a hydroelectric project.

Construction

Dam building is a hard and skilled job which begins with diverting the flow of the water temporarily. Following this, the area of the river bed which acts as the site for the foundation is prepared for construction, which involves ensuring that the rocky or sedimentary bed is strong enough to support the dam. The area on either side may also be strengthened. Following this a concrete block formation is built up. Once it is high enough, the water is allowed to flow in under strict control with careful monitoring. Other operational structures are finally added.

How it Works

Dams are the high-visibility part of a hydroelectric system, storing water at a height. When this water is released it flows down by gravity, and the potential energy is converted to kinetic energy. The height of the water level and the rate of flow of water decide the electricity output. The water flow is regulated by a gate or shutter which in turn is determined by the demand for power. The water flows down a series of channels or penstocks which guide and regulate the steepness of the fall, so as to optimize the efficiency of the dam. The water finally passes through a hydraulic turbine and then into a tailrace, or exit stream, at the bottom of the dam, to rejoin the river.

Hydroelectric power is generated by the controlled flow of water through the turbine, a device that has a shaft connected to a generator that contains metal coils surrounded by magnets. As the water spins the turbine shaft, the magnets revolve around the metal coils, leading to electromagnetic induction, a phenomenon which generates electric power. In this way potential energy in the stored water is converted to kinetic energy as it flows down the penstock, and then into mechanical energy in the rotating turbine shaft, which in turn is converted to electrical energy.

Energy demands vary over the day. One advantage of hydroelectric plants is that they can vary their output to cater to peak power demands over short periods by the incorporation of pumped storage. The water which flows through the turbine is pumped into a higher storage pool located above the turbine, using excess power generated during the period of low consumer demand for power. At peak power demand this is released into the lower stream again to generate more electricity. In this way the reservoir behaves as a type of battery.

Advantages

Hydroelectric plants can be put into operation quickly. Their power output is easily adjusted. Their operation remains efficient whether for a short or long period. Construction costs are low for small or medium-sized dams compared with other hydropower plants. A dam produces electricity without burning fossil fuels, and is thus considered a source of clean energy.

Sizes

Hydroelectric dams come in a range of sizes:

- Large: A dam which has a capacity of over 30 MW is a large project.
- Small: A hydroelectric dam with a capacity of 10 MW or less.
- Micro: A microhydroelectric dam generates up to 100 KW of power, such as for a single home, a ranch or a village.

Disadvantages

Dams do produce some negative environmental impact. For instance, many species of fish cannot migrate for spawning purposes once their rivers are blocked by dams. In the USA alone, some salmon populations have come down to a couple of million from over 15 million following dam construction.

Dams and reservoirs also cause the natural temperature, the chemical composition, the type of flow and the silt loads in the river water to change, which affects the life in the river and its banks.

Dam reservoirs spread over hundreds or millions of hectares of agricultural land or land that have been settled on by a large and productive population, or archeological sites.

The human cost of dam building includes these physical changes, resettlement, and other far-reaching consequences. The concrete and other materials used in dam construction are products of fossil fuel burning, but this one-time environmental cost is thought to be offset by the availability of clean power over a long period of 50 to 100 years.

Sometimes reservoirs serve as a source of greenhouse gases such as carbon dioxide and methane, which can have significant impact due to the large amount produced. However, scientists have developed means to use this methane to produce more electricity, which can enhance the power output while minimizing the environmental cost.

Types of Dams

Various types of dams are constructed based on their usage. These dams are classified based on construction material used, structure and design, functions, hydraulic design and size.

Concrete Dams

Concrete dams are built in four basic shapes. The concrete gravity dam has weight as its strength. A cross section of this dam looks like a triangle, and the wide base is about three-fourths of the height of the dam. Water in the reservoir upstream of the dam pushes horizontally against the dam, and the weight of the gravity dam pushes downward to counteract the water pressure. The concrete buttress dam also uses its weight to resist the water force. However, it is narrower and has buttresses at the base or toe of the dam on the downstream side. These buttresses may be narrow walls extending out from the face of the dam, much like the "flying buttresses" supporting cathedral walls or a single buttress rather like a short dam may be built along the width of the toe of the dam.

The arch dam is one of the most elegant of civil engineering structures. In cross section, the dam is narrow in width, but, when viewed from above, it is curved so the arch faces the water and the bowl of the curve looks downstream. This design uses the properties of concrete as its strength. Concrete is not strong in tension (when it is pulled or stretched), but it is very strong in compression (when it is pushed or weighed down). The arch dam uses the weight of the water behind it to push against the concrete and close any joints; the force of the water is part of the design of the dam. The arch-gravity dam is a combination of the arch type and gravity type, as the name suggests; it is a wider arch shape. Multiple-arch dams combine the technology of arch and buttress designs with a number of single arches supported by buttresses.

Concrete dams are used more often than fill dams to produce hydroelectric power because gates (also called sluices) or other kinds of outlet structures can be built into the concrete to allow for water to be released from the reservoir in a controlled manner. When water for power, drinking water, or irrigation is needed downstream, the gates can be opened to release the amount needed over a specified time. Water can be kept flowing in the river downstream so fish and other wildlife can survive. Both concrete and fill dams are required to have emergency spillways so that flood waters can be safely released downstream before the water flows over the top or crest of the dam and potentially erodes it. Spillways channel the water downstream and well below the base or toe of the dam so the dam and its foundation are not eroded.

Most dams built in the twentieth century and those being designed today have several purposes. Over 40,000 dams higher than 45 ft. (15 m) and classified as large dams exist, and more than half of these have been built since 1960. Of these dams, 16% of them are in the United States and 52% are in China; 83% are fill dams used primarily for water storage, and the remaining 17% are concrete or masonry dams with multiple purposes. Dams that generate hydroelectric power produce 20% of the electricity in the world.

Raw Materials

The key raw materials for concrete dams are concrete itself and steel reinforcement. number of other materials and components made by specialty contractors may be used in dam building and include steel gates and tunnel liners, rubber water stops, plastic joint-filling compounds to prohibit the movement of water, electrical controls and wiring, siphons, valves, power generators, a wide assortment of instruments, and even Teflon sheeting to line water outlet structures to prevent turbulence and cavitation (damage due to swirling water).

Concrete it is made of cement, water, and materials collectively called aggregate that consist of sand or gravel. Cement has unique properties that must be considered in selecting the cement, designing the dam, and timing construction. Mixing of cement and water causes a chemical reaction that makes concrete hard but that also releases heat. This causes a distinct rise in the temperature inside a mass of concrete, and, when the concrete begins to cool, it shrinks and cracks, potentially causing leaks. To limit these effects, concrete can be placed when the air temperature is low, low-heat cement can be used, and water can be circulated through pipes in the concrete. Furthermore, the concrete has to be placed in shallow lifts (i.e., only a few feet or meters are added at a time) and in narrow blocks; then it has to be allowed to cure over a specified minimum time so the heat dissipates. Depending on the design of the dam, engineers will choose the concrete mix (including the cement and type of aggregate) very carefully; a thin arch dam is designed with a different concrete mix than a massive gravity dam.

Design

Design of a concrete dam depends on the purpose of the dam and the configuration of the site where it will be built. Dams are of two general types. Overflow dams block flow in a stream and harness the water for generating power or to improve navigation and provide irrigation water. The components of an overflow dam are designed so the water can be released and the level of the water in the reservoir regulated by a series of sluice gates, spillways, or outlet tunnels. Non-overflow dams store water for drinking water supply, irrigation, or power; they also have a spillway, but its use is restricted for emergencies to lower the water level quickly during floods. Methods for releasing the stored water are much more limited than in overflow dams, and the dam itself may not contain any outlet structures. Instead, water may be pumped out for irrigation, for example, from part of the reservoir.

Some sites are best suited to particular types of dams. An arch dam is most appropriate for construction in a high, narrow gorge where the arch of the structural shape provides strength. But an arch can also be built across a wider canyon where other effects like friction on the base of the dam add strength and resistance to movement. Similarly, a gravity dam is the typical choice for a shallow, wide canyon, but if it is built with some curvature, arching action will also strengthen a gravity dam in a narrower and higher gorge. Where the riverbed is exceptionally wide, the dam may be designed to have several spans, each with different engineering properties depending on the variation of foundation materials. The separate spans are usually supported on the downstream (air) side by buttresses or the extended curves of multiple arches. Sometimes, the spans of multiple span dams are constructed of concrete slabs or steel plates supported on piers.

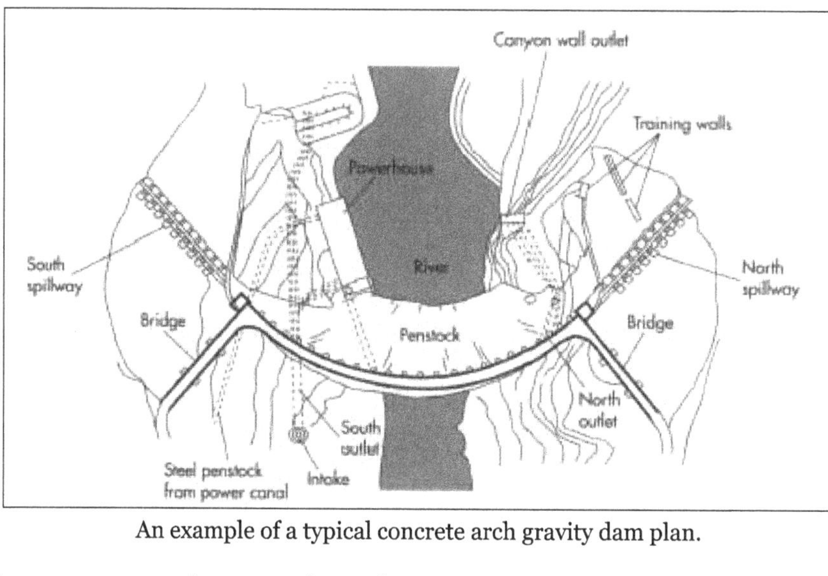

An example of a typical concrete arch gravity dam plan.

Like fill dams, concrete dams go through extensive rounds of preliminary design and feasibility studies to choose and explore the site, to evaluate the quantity of water retained and its value (as a power source or source of supply) versus the cost of the project over the anticipated years of operation, to consider a wide range of other effects such as changes to the environment, and to choose a dam of the optimal size and configuration. Hundreds of factors enter into these studies, and the process is usually iterative. A design is chosen and tested against all these factors until it fails to satisfy one or more factors, and the next variation in design is chosen and studied until it fails—or passes.

The design process for a concrete dam typically involves professionals from a more extensive range of disciplines than design of a fill dam. The technical professionals who contribute their expertise to design of a concrete dam may include geologists, seismologists, environmental scientists, geotechnical (soil) engineers, civil engineers, structural engineers, computer analysts (specialists in software applications that examine the dam's strength and safety), hydrologists and hydraulic engineers, mechanical engineers, and electrical engineers if the dam is to be used for power generation. Still more specialists may study aspects like corrosion of concrete and steel structures. The teamwork required for dam design and construction is critical not only because of the enormous costs of these projects but because the safety of persons and property downstream demands perfection.

The Construction Process

1. Before construction can begin on any dam, the water in the streambed must be diverted or stopped from flowing through the site. As in the case of fill dams, a coffer-dam (a temporary structure to impound the water) must be built or the water must be diverted into another channel or area down-stream from the dam site. For large projects, this construction may be done several seasons before building of the dam begins. The flow of water is closed off at the very last moment.

2. The foundation area for any concrete dam must be immaculate before the first concrete for the dam is placed. As for fill dams, this is a detailed process of excavating, cleaning, and repairing the rock throughout the foundation "footprint" and on both abutments (the sides of the canyon that form the ends of the dam). Sites immediately downstream of the dam for any power plant, stilling basin, or other structure must also be prepared.

At some sites, extensive work may be required. If the rock in the foundation or abutments is prone to fracturing because of the load imposed by the dam and its reservoir, earthquake activity, or the properties of the rock, it may be necessary to install extensive systems of rock bolts or anchor bolts that are grouted into the rock through potential fracture zones. On the abutments above the dam, systems of rock bolts and netting may be required to keep large rock fragments from falling onto the dam. Instruments to monitor groundwater levels, joint movement, potential seepage, slope movements, and seismic activity are installed beginning during the early stages of foundation preparation through completion of the dam.

A cutoff wall may be excavated deep into rock or holes may be drilled in the foundation for the installation of reinforcing steel, called rebars that extend up into the dam and will be tied to the steel inside the first lifts of the dam. The idea is to build a reservoir that, like a bowl, is equally sound around its perimeter. The water is deepest and heaviest at the dam (when the reservoir is near capacity) so the dam and its foundation cannot be a weak point in that perimeter.

3. Forms made of wood or steel are constructed along the edges of each section of the dam. Rebar is placed inside the forms and tied to any adjacent rebar that was previously installed. The concrete is then poured or pumped in. The height of each lift of concrete is typically only 5-10 ft. (1.5-3 m) and the length and width of each dam section to be poured as a unit is only about 50 ft. (15 m). Construction continues in this way as the dam is raised section by section and lift by lift. Some major dams are built in sections called blocks with keys or inter-locks that link adjacent blocks as well as structural steel connections.

The process is much like constructing a building except that the dam has far less internal space; surprisingly, however, major concrete dams have observation galleries at various levels so the condition of the inside of the dam can be observed for seepage and movement. Inlet and outlet tunnels or other structures also pass through concrete dams, making them very different from fill dams that have as few structures penetrating the mass of the dam as possible.

4. As soon as a significant portion of the dam is built, the process of filling the reservoir may begin. This is done in a highly controlled manner to evaluate the stresses on the dam and observe its early performance. A temporary emergency spillway is constructed if dam building takes more than one construction season; lengthy construction is usually done in phases called stages, but each stage is fully complete in itself and is an operational dam. The upstream cofferdam may be left in place as a temporary precaution, but it is not usually designed to hold more than minimal stream flows and rainfall and will be dismantled as soon as practical. Depending on design, some dams are not filled until construction is essentially complete.

5. The other structures that make the dam operational are added as soon as the elevation of their location is reached as the dam rises. The final components are erosion protection on the upstream (water) side of the dam (and sometimes downstream at the bases of outlet structures), instruments along the crest (top) of the dam, and roads, side-walks, streetlights, and retaining walls. A major dam like Hoover Dam has a full-fledged roadway along its crest; small dams will have maintenance roads that allow single-file access of vehicles only.

Away from the dam itself, the powerhouse, instrument buildings, and even homes for resident operators of the dam are also finished. Initial tests of all the facilities of the dam are performed.

6. The final details of constructions are wrapped up as the dam is put into service. The beginning of the dam's working life was also carefully scheduled as a design item, so that water is available in the reservoir as soon as the supply system is ready to pump and pipe it downstream, for example. A program of operations, routine maintenance, rehabilitation, safety checks, instrument monitoring, and detailed observation will continue and is mandated by law as long as the dam exists.

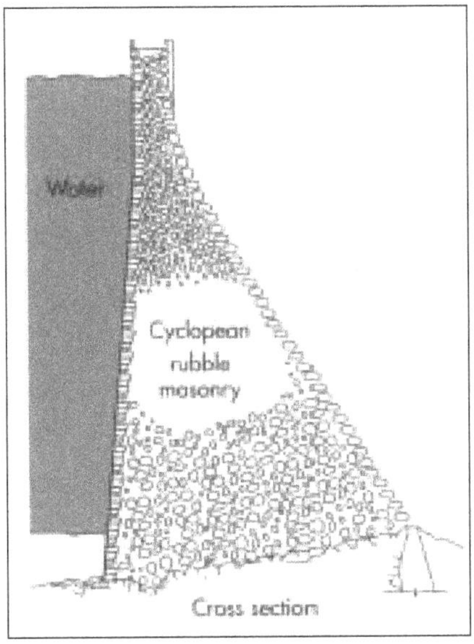

Cross section of a typical concrete arch gravity dam. The height is 280 ft. (85 m). The thickness grows from 16 ft (4.9 m) at the top to 184 ft. (56 m) at the base.

Quality Control

There is no dam construction without intensive quality control. The process of building alone involves heavy equipment and dangerous conditions for construction workers as well as the public. The population living downstream of the dam has to be protected over the structure itself; the professionals who design and construct these projects are absolutely committed to safety, and they are monitored by local, state, and federal agencies like Divisions of Dam Safety, the U.S. Corps of Engineers, and the Department of Reclamation.

Byproducts/Waste

There are no byproducts in dam design or construction although a number of other associated or support facilities may be needed to make the project work. Waste is also minimal because materials are too expensive for waste to be allowed. Also, locations are often remote, and the process of hauling waste away from the site and disposing it is prohibitive. Soil and rock that may be excavated from the foundation area, down-stream sites, the abutments, or portions of the reservoir are usually used elsewhere on the project site. Quantities of materials cut away or placed as fill are carefully calculated to balance.

The Future

The future of concrete dams is the subject of much debate. Each year, over 100,000 lives are lost in floods, and flood control is a major reason for building dams, as well as protecting estuaries against flooding tides and improving navigation. Lives are also benefited by dams because they provide water supplies for irrigating fields and for drinking water, and hydroelectric power is a non-polluting source of electricity. Reservoirs are also enjoyed for recreation, tourism, and fisheries.

However, dams are also damaging to the environment. They can change ecosystems, drown forests and wildlife (including endangered species), change water quality and sedimentation patterns, cause loss of agricultural lands and fertile soil, regulate river flows, spread disease (by creating large reservoirs that are home to disease-bearing insects), and perhaps even affect climate. There are also adverse social effects because human populations are displaced and not satisfactorily resettled.

Timber Dams

Timber dams are slightly permeable gully blocks that are used both to slow the flow of water downstream, and to create pools. They are (deliberately) leaky, such that they can trap water from high rainfall events and then slowly release some or all of it (thereby making an empty space ready to trap the next rainfall event). During this process they also trap sediment that would otherwise be lost from the moor. Slowly the sediment accumulates until it fills the pool, at which time the dam loses its water storage function but has raised the bed of the gully, which can then be recolonized by vegetation such as cotton grass or Sphagnum moss (possibly artificially aided, e.g. through plug planting or seeding).

Dams should be no more than five or six planks high. The planks and posts must be driven into the peat sufficiently to prevent undercutting of the peat, and should be keyed into the sides by at least 30cm to prevent side-cutting. A splash plate of stone or timber should be positioned beneath the spillway as a baffle to prevent undercutting.

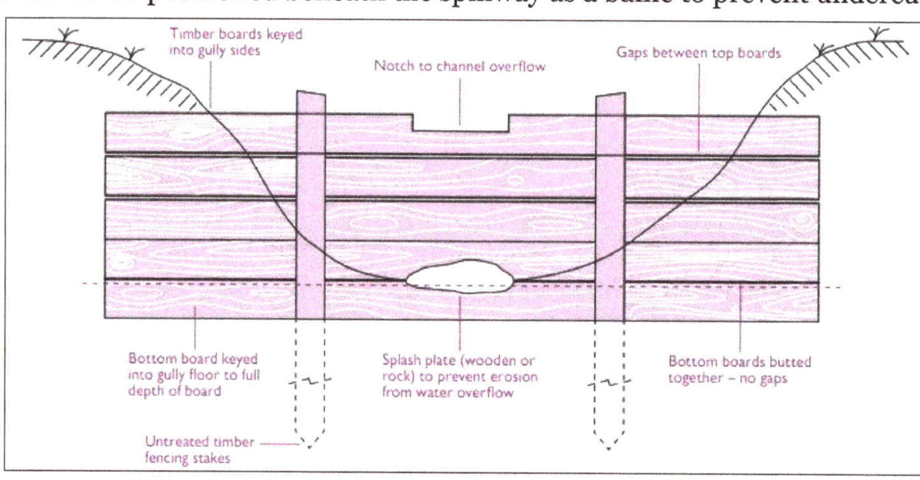

Timber dam construction.

Purpose:

- Can create pools, and trap sediment, allowing for re-vegetation.

Advantages:

- No vehicular access required – timber can be airlifted to any location and dams constructed using hand tools.
- Can be installed across a range of gully profiles.

Disadvantages:

- Require enough peat in which to drive stakes to support dam.
- Can make it difficult to cross gullies on foot when pools are full.

Gravity Dams

A Gravity dam is defined as a solid structure, made of concrete or masonry, constructed across a river to create a reservoir on its upstream. The section of the gravity dam is approximately triangular in shape with its apex at its top and the maximum width at the bottom. The section is so proportioned that it resists the external force acting on it by its own weight.

When this type of structure is most durable and requires very little maintenance. Nowadays, concrete gravity dams prefer as compared to other dams. They can be constructed with ease on any dam site, where a natural foundation strong enough to bear the enormous weight of the dam is available.

Construction of Gravity Dam

Foundation

A suitable foundation for a gravity dam such as a sound rock foundation should be available.

The Topography of Gravity Dam

The river-valley at the dam site should preferably have a narrow gorge to reduce the dam. However, the gorge should be open out upstream to provide a large basin for a reservoir. The general bed level at the dam site should preferably be higher than that of the river basin. This will reduce the height of the dam and will facilitate the drainage problem.

Materials

Materials required for dam construction should be available nearby, to avoid much transportation. When this would very much reduce to cost of construction.

Reservoir and Catch Area

- The cost of land and property submerged by the proposed dam should be as minimum as possible.
- The site should have an adequate storage capacity of the reservoir basin at a minimum cost.
- The reservoir basin should be reasonably water-tight. The stored water should bot escape out through its sidewalls and bed.
- The geological condition of the catchment area should such that the percolation losses are minimum and maximum run-off obtain.

Communication

The dam site should easily accessible to get connected by road, rail, etc. For transportation of cement, labor, and other equipment.

Locality

The surroundings near the site should be healthy and free of mosquitoes etc. As staff and labor colonies have to be constructed near the site.

Forces Acting on a Gravity Dam

- Weight of the dam,
- Water pressure,
- Uplift pressure,
- Silt pressure,
- Wind pressure,
- Water pressure,
- Wave pressure,
- Ice pressure,
- Earthquake forces.

Weight of the Dam

The weight of the dam is the main stabilizing force in the gravity dam. Generally, for analysis purposes, the unit width of the dam considers. When the cross-section of the dam may divide into several rectangles and triangles, and the weights w1, w2, and w3, etc. of each of these can be determined, along with the determination of their lines of action. The total weight of the dam acts at the center of gravity of its section.

The Weight of the Damper Unit Length

Product of area of the dam X Specific weight of the material.

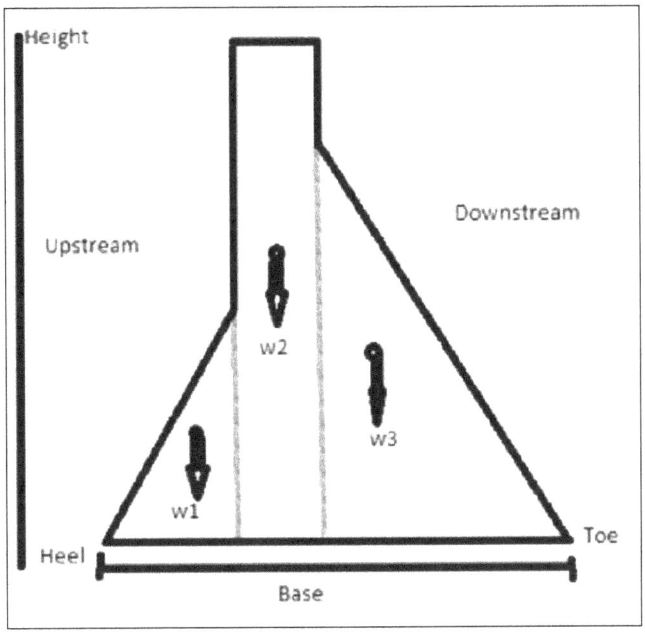

Weight of Gravity dam.

Ice Pressure

The ice which is formed on the water surface of the reservoir in the cold countries may sometimes melt and expand. The upstream face of the reservoir then has to resist the thrust exerted by expanding ice. This force act linearly along the length of the dam and at the reservoir level. When the magnitude of these forces may vary from 250 to 1500 kN/m^3 depending upon the temperature variation. On average, a value of 500 kN/m^3 can allow under ordinary conditions.

Wind Pressure

Wind pressure acts on the exposed surface of the dam when winds blow. Usually, the wind pressure is not significant for the design of gravity dams and is therefore neglect. However, the superstructure should consider in the design. A wind pressure of 1 to 1.5 kN/m^2 generally consider over the area exposed to high winds.

Earthquake Forces

If the dam is located in a seismic region, it should be designed for earthquake forces. An earthquake produced seismic waves which are capable of shaking the earth upon which the dam is resting, and sets the dam into a state of vibration. When the seismic waves originate from a point known as the focus of the earthquake, which is located deep into the earth. The corresponding points on the earth surface which lie vertically above the focus know an epicenter.

The seismic waves impart a momentary acceleration in the earth's crust which moves in the direction in which the waves traveling at the instant. If the ground under the dam moves, the dam must also move with it to avoid rupture. Inertia forces under consideration in the design of the dam due to the earthquake.

Advantages of Gravity Dam

- A gravity dam requires the least maintenance than other dams.
- Gravity dams are relatively more strong and stable as compared to earth dams.
- It can build any high whereas the high of an earth dam.
- Gravity dam does no trial suddenly; it gives enough warning time whereas earth dams generally fail suddenly.
- Deep-set sluices can be used in gravity dam, to retard the sedimentation or silt deposit in the reservoir.

Disadvantages of Gravity Dam

- Gravity dams can be built only on a sound rock foundation. While earth dams can be built on weak foundations or permeable foundations with appropriate foundation treatment.
- The initial cost of a gravity dam is always higher as compared to an earth dam.
- If mechanized plants, such as manufacturing and transporting mass concrete, curing of concrete, etc. are not available, a gravity dam may take more time to construct.
- Gravity dams require skilled labor for is construction.

Earth Dams

Earth dam utilizes natural materials with a minimum of processing and may be built with primitive equipment under conditions where any other construction material would be impracticable. It is not surprising that the earliest known dams were of earth.

Modern developments in earth-moving equipment have resulted in decreased cost for earth moving as compared with an increase in the cost of concrete as a result of increased wage and material costs.

Earth dams are now competitive in cost with masonry in all sizes. The design of earth levees and dikes follows the same principles as for earth dams. Unlike high-arch and gravity dams, which require a sound rock foundation, earth dam is readily adapted to earth foundations.

They are logical choices for many sites where foundation conditions make concrete dams unsatisfactory. It should not be assumed that the construction of earth dam is a simple operation and that their design requires little more than rule-of-thumb criteria.

Numerous failures of poorly designed earth embankments make it apparent that earth dams require as much engineering skill in their conception and construction as any other type of dam. In no other type of dam are the construction and design procedures so interdependent; continuous field observations of deformations and pore water pressures are often made during the construction period to evaluate the initial design. Modifications of design based on these observations are not uncommon for large earth dams.

Types of Earth Dams

The simple embankment is essentially homogeneous throughout, although a blanket of relatively impervious material may be placed on the upstream face. Levees are often simple embankments, but large dams are rarely constructed in this manner.

Zoned embankments usually have a central zone of selected soil material, to form a relatively impermeable core, a transition zone along both faces of the core to prevent piping through cracks which may form in the core, and outer zones of more perviousness material for stability.

This construction is widely used in earth dam and is selected whenever suitable materials are available. Clay, even though highly impermeable, may not make the best core if it shrinks and swells too much. The most satisfactory cores are of clay mixed with sand and tine gravel.

Diaphragm-type dams have a thin central section of concrete, steel, or timber which serves as a water barrier, while the surrounding earth or rock till provides stability. Thin concrete sections are easily cracked by differential earth loads, and it is difficult to form a perfectly watertight barrier of timber or steel.

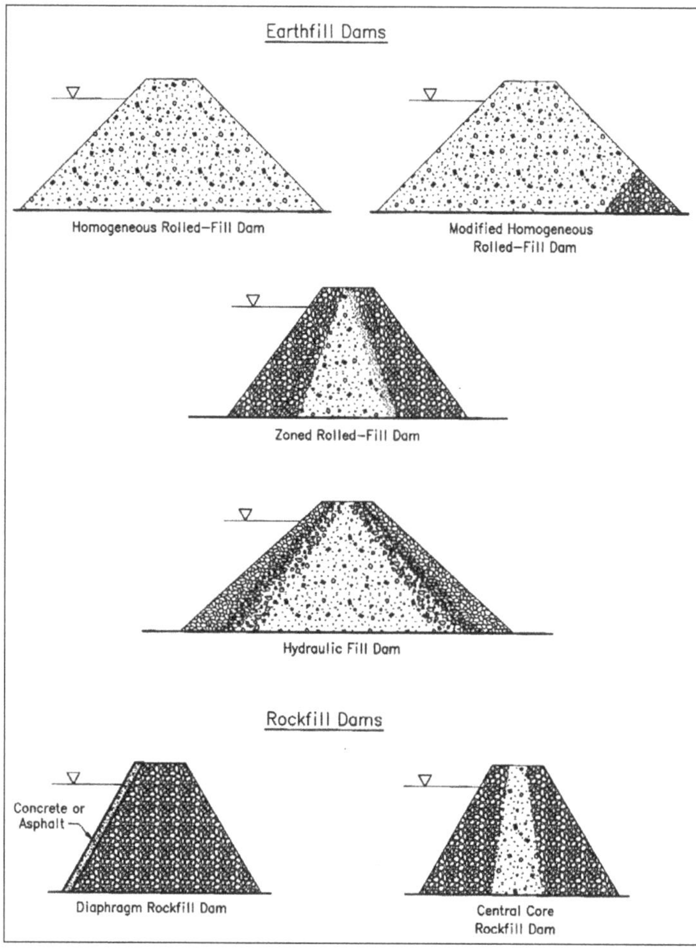

Schematic of the major types of embankment dams.

In addition, the diaphragm must be tied into bedrock or a very impermeable material if excessive under-seepage is to be avoided. Rock fill dams are built of coarse rock, which provides structural stability, with a concrete membrane on the upstream slope as a water barrier.

Methods of Construction of Earth Dam

Hydraulic-fill dams are constructed by using water for transporting the material to its final position in the dam. The material is discharged from pipes along the outside edges of the fill, and the coarse materials are deposited soon after discharge, while the lines are carried into the central pool. The result is a zoned embankment with a relatively impermeable core.

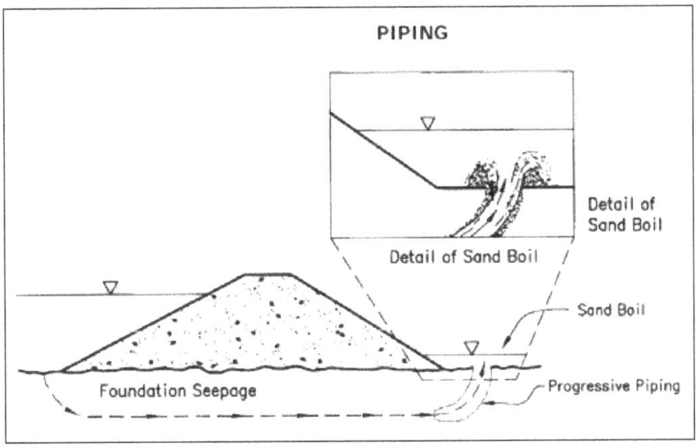

This type of construction originated in the western United States as a result of experience with hydraulic mining. Recent improvements in equipment for transporting and compacting dry fill have tended to discourage use of the hydraulic method.

Hydraulic fill is best suited for placing well-graded soils containing a considerable amount of coarse sand and gravel. Adequate water supply is necessary, although some of the water may be re-circulated.

Since the fill is saturated when placed, high pore pressures develop in the core material, and the embankment must be designed for stability under these pressures. Because of the slow drainage of the water from the core, a considerable settlement is to be expected over a long period.

The embankment is especially susceptible to earthquake damage until drainage is reasonably complete. Semi-hydraulic-fill dams are constructed by dumping the fill material from trucks into its approximate position in the dam and sluicing the tines into a core. Like the hydraulic-fill method, this procedure requires careful control to assure a satisfactory embankment.

Dumping soil in a pool of water to produce a puddled core often washes the fines from the material and leaves a more permeable core than might have been obtained from proper dry compaction. Numerous failures of hydraulic fill dams have occurred during construction.

Rolled-fill dams are constructed by placing selected materials in thin layers (6 to 18 inches), and compacting them with a heavy roller. Some compaction may be obtained by proper routing of trucks and other construction equipment. Usually, however, special equipment is used for compacting the fill. Both sheep-foot rollers and heavy pneumatic-tired rollers are used singly or in combination.

Ordinary road rollers have been used successfully on small projects. In any case, the material should be placed at moisture content near that for optimum density. Coarse gravels are not suited for compaction by rolling, but vibrating equipment can be used.

Rock-fill Dams

Rock-fill dams are a kind of dam that combines porous yet firmly packed gravel, sand, or silt in one section with an area of concrete, metal, clay, or some other substance that completely blocks water. The section of compacted granular soil helps slow the flow of water while the impervious zone provides strength and stability. Rock-fill dams are one of two types of embankment dams, the other being earth dams.

These structures are effective because the force of the river or reservoir hits the impervious zone and is transferred to the firmly packed transition zone where water can slowly begin to filter through the dam. When clay is used for the impervious zone, this is known as a composite dam.

Rock-fill Components

Rock-fill dams are made of two main parts, the main rock-fill or transition zone and the impervious zone. Synthetic filters made of specifically graded soil are also sometimes used to reduce internal erosion (that is, to stop fine grains from moving around within the dam).

Special Considerations

If you're ever assigned to build such a dam or you decide that you want to build a rock-fill dam on your property, it's important that you pay attention to the following points:

- Stability is critical. Both the foundation and the abutment must be stable throughout the life cycle of the dam.
- Seepage must be controlled to ensure safe operation. Seepage control measures should be in place before the dam's initial use.
- The spillway must be strong enough to prevent overtopping of the embankment.
- Reservoir slopes should be carefully designed.

Advantages of Rock-fill Dams

One of the most important aspects of a rock-fill dam is that it can be very cheap to build since materials can be sourced from near the dam's location. This type of dam can also be suitable in areas that don't present the best conditions for deep foundations. Another key benefit is that the dam can be planned in areas that are susceptible to cold conditions, and in hot and humid climates, as well. Moreover, they can be built even in rough weather, and they take far less time to build than many other types of dams.

How to Create a Rock-fill Dam

The best way to build a dam is to start by placing the material in controlled lifts. Steel-drum roller compactors are used so the material can be locked into place to reduce settlement during subsequent years. These rollers can also be used to avoid material segregation due to high lift rock fall on the slopes.

The dam height can be naturally increased by adding more rocks behind the diaphragm without encroaching on the reservoir area. The main rock-fill provides support for its own weight and stability, while the membrane holding the water creates an impervious zone. This membrane can be made of wood, concrete, steel, asphalt, or stone. The dam is then made water-tight by continuing the impervious face upward. Always consult your local building codes and water regulations before proceeding.

Arch Dam

An arch dam is a dam curved in plan and carries at major part of its water load horizontally to the abutments by arch action. This part of water load depends primarily upon the amount of curvature. The balance of the water load is transferred to the foundation by cantilever action. The thrust developed by the water load carried by arch action essentially requires strong side walls of the canyon to resist the arch forces.

The weight of arch dams is not counted on to assist materially in the resistance of external loads. For this reason, uplift on the base is not an important design factor. Many early arch dams were built of rubble, ashlar masonry, etc. However, practically all arch dams constructed during recent years have been built of concrete.

The famous Hoover dam of U.S.A., constructed in 1936, is an an arch dam of constant radius type having the following dimensions:

- Height 222 m (726 ft.)
- Length 380 m (1244 ft.)

- Top width 13.7 m (45 ft.)
- Base thickness 201 m (660 ft.)

Advantages of Arch Dam

- Arch dams are particularly adapted to the gorges where the length is small in proportion to the height.
- For a given height, the section of an arch dam is much lesser than a corresponding gravity dam. Hence, an arch dam requires less material and is, therefore, cheaper.
- Because of much less base width, the problems of uplift pressure are minor.
- Since only a small part of water load is transferred to the foundation by cantilever action, an arch dam can be constructed in moderate foundations where gravity dam requiring sound foundation rock may be unsuitable.

Disadvantages of Arch Dam

- It requires skilled labor and sophisticated form work. The design of an arch dams are also quite specialized
- The speed of construction is normally slow.
- It requires very strong abutments of solid rock capable of resisting arch thrust.

Hence, it is not suitable in the locations where strong abutments are not available. Unfortunately, only few sites are suitable for this type of dam.

Types of Arch Dams

The arch dams may be divided into the following three types:

Constant Radius Arch Dam

A constant radius arch dam is the one in which the radius to some prominent feature (such as upstream face, or the downstream face, or the arch axis) is constant. In most cases, however, the constant radius refers to the upstream side.

In a constant radius arch dams, the arch centers for upstream face, downstream face and the center line of the horizontal arch rings at various elevations lie on a straight vertical line; that passes through the center of the horizontal arch ring of the upstream face at the crest. Hence, this type of arch dam is also sometimes known as the constant center arch dam.

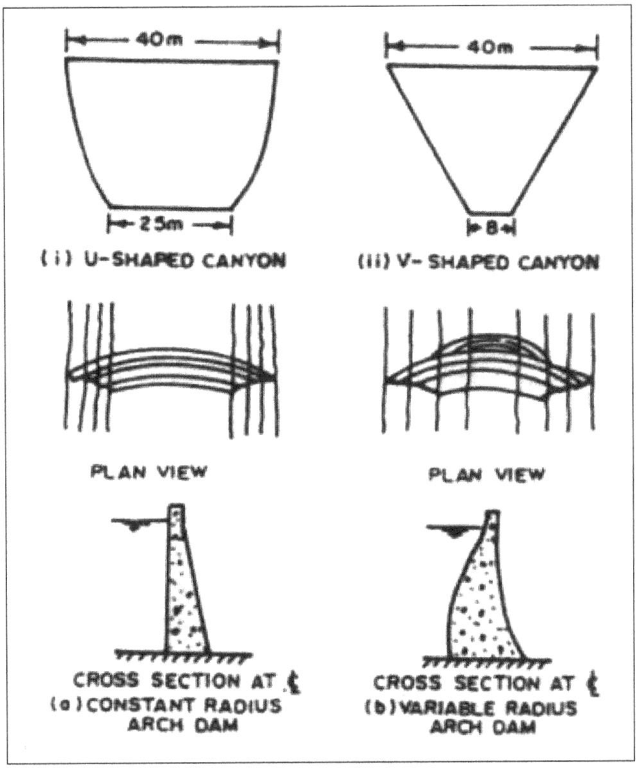

Constant and variable radius arch dams.

In most of the simple arch types, the upstream face is vertical (i.e., its radius is constant at all elevations), and the thickness is proportional to the height. This makes the central angles of the arch rings of the introdos vary at various elevations, the maximum being at the top of the dam and a certain minimum at the bottom of the dam.

A constant radius arch dam is particularly adapted to U-shaped canyons as cantilever action will carry a large proportion of the load at lower levels. The constant radius arch dams are less economical than the constant angle arch dams, and the latter is constructed in preference to constant radius arch darn. However, the form work for the constant radius arch dam is much simpler to construct.

Variable Radius Arch Dam

A variable radius arch dam is the one in which the radii of the extrodos curves (i.e., arch rings corresponding to upstream face); and introdos curves (i.e., arch rings corresponding to downstream face) vary at various elevations being maximum at the top and certain minimum at its bottom.

This makes the central angle as large as possible so that maximum arch efficiency may be obtained at all elevations. The variable radius arch dams have often vertical or even overhanging laces at the upstream side near the abutments and the downstream side near the crown. The centers of the horizontal arch rings at various elevations do not lie on the single vertical line.

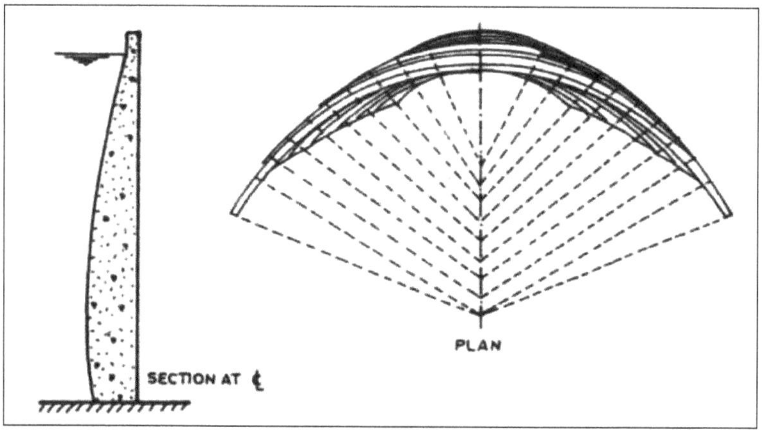

Variable radius of Arch dam.

Hence, it is also known as variable center arch dam. A variable radius arch dams are often adapted to narrow V-shaped canyons since arch action can be depended upon at the lower levels. A variable radius arch dam has greater arch efficiency, resulting in saving of concrete. Hence, it is preferred in comparison to the constant radius arch dam.

Constant Angle Arch Dams

The constant angle arch dam is a special type of variable radius arch dam in which the central angle of the horizontal arch rings is of the same magnitude at all elevations.

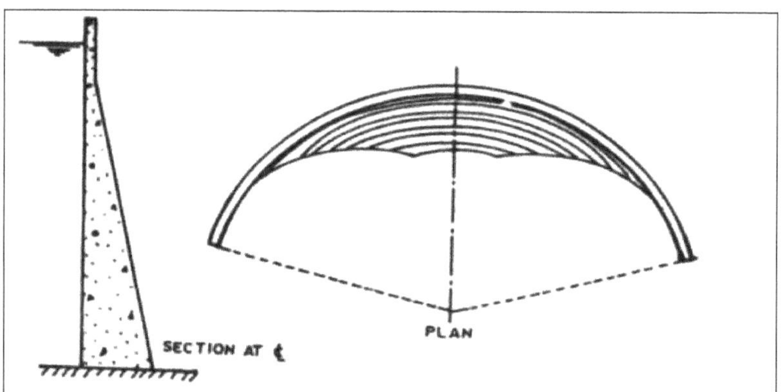

Constant radius of Arch dam.

It has been shown that the volume of concrete is minimum when the central angle is 133° 34′. A variable radius arch uses about 58% of the concrete required in constant radius arch dams. A constant angle arch dams requires about 42.6% of the concrete required in constant radius arch dams. Hence, a constant angle arch dam is the most economical.

Buttress Dam

A buttress dam consists of water retaining sloping membrane or deck on the upstream which is supported by a series of buttresses at right angles to the axis of the dam. The upstream water retaining sloping membrane may be in the form of a reinforced concrete slab or a series of arches or thickened buttress heads. At the upstream end a cutoff is provides to prevent or reduce the seepage of water.

Buttresses are the thin walls of triangular profile shape with a characteristic sloping upstream face. They are usually spaced at equal intervals along the length of the dam and are supported either on a continuous mat foundation or on a separate spread footing.

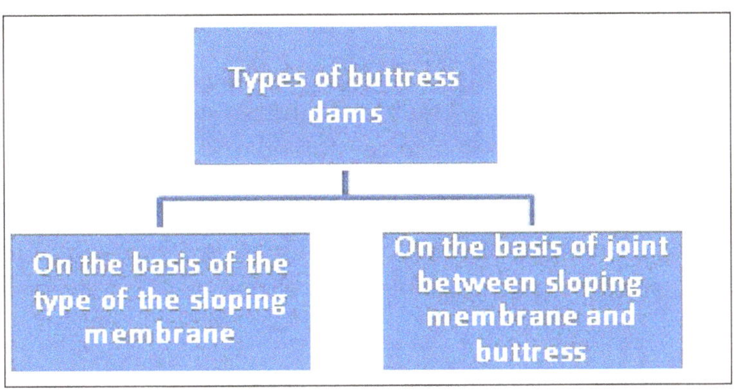

Classification based on the Type of Sloping Membrane

On the basis of sloping membrane the buttress dams may be classified as follows:

Ambursen Type/Flat Slab/Deck Slab Type Buttress Dams

The deck slab or flat slab type buttress dams are also known as Ambursen type buttress dams in honor of Nils F Ambursen who in1903 built the first flat slab type of buttress dam. In this type of buttress dam the sloping membrane or deck consists of a reinforced concrete slab supported by a series of buttresses. The inclination of the deck slab is kept between 35? to 45? with the horizontal. In order to provide a wide base for the slab supported by the buttress the upstream end of the buttress where it joins the slab is usually made wide by providing haunch or corbel. The deck slab type buttress dams can be subdivided into three types:

- Simple deck slab type buttress dams: In this type of buttress dam the deck slab is not rigidly attached to the buttresses but it is in the form of a simply supported slab which remains in contact with the buttress tongue. The joint between the slab and buttress tongue is filled with bituminous mastic or asphaltic putty or some flexible joint compound.
- Fixed or continuous deck slab type buttress dams: In this type of buttress dam the deck slab is cast monolithic with the buttress and it is in the form of a continuous slab. Thus in this case the reinforcement is provided both on upstream and downstream faces of the slab because of reversing bending moments and there is a possibility of upstream reinforcement being corroded. In this type of dam the deck slab is relatively thin.
- Cantilever deck slab type buttress dams: In this type of buttress dam the deck slab is cast monolithic with the buttress in such a manner that the deck slab overhangs on either side of each buttress as cantilever.

In addition to the above noted deck slab type buttress dams, there are two more types of these dams as following:

- Columnar type buttress dams: This type of dams consist of continuous flat slab of reinforced concrete supported on a number of inclined columns instead of buttresses. The columns are also reinforced and are stiffened by a two-directional system of struts.
- Truss type buttress dams: IN this type of buttress dams the reinforced-concrete deck slab is supported on a number of reinforced-concrete trusses instead of buttresses.

Multiple-arch Type Buttress Dams

In this type of buttress dam the sloping membrane or deck consists of a series of reinforced-concrete arches supported by a number of buttresses. The upstream face of the dam is usually inclined at 45.The arches are cast monolithic with the buttresses.

Multiple-dome Type Buttress Dams

In this type of buttress dam the sloping membrane or deck consists of a series of reinforced-concrete domes supported by a number of buttresses.

Bulk Head/Massive Head Type Buttress Dams

The main feature of this type of dam is, there is no separate water retaining member is provided and the water retaining member is formed only by enlarging the upstream end side of the buttress. Thus the dam is made with a series of buttresses and massive heads placed side by side. The massive head type buttress dams may be further subdivided on the basis of the shape of the buttress head into the following 3 types:

- Round head buttress dams.
- Diamond head buttress dams.
- Tee head buttress dams.

Classification based on the Joint Between the Sloping Membrane and Buttresses

On the basis of the joint between the sloping membrane and buttresses the buttress dams may be classified in the following three categories are:

Rigid Buttress Dams

Here, the sloping membrane or deck is constructed monolithic with the buttresses. As such for these dams the influence of foundation settlement, temperature variation and shrinkage of concrete are significant and provisions have to be made for these in the design.

Articulated Buttress Dams

Here, the sloping membrane or deck is not constructed monolithic with the buttresses. As such dams are flexible and hence the influence of settlement of foundation, temperature variation and shrinkage of concrete are not significant for these dams.

Intermediate or Semi-rigid Buttress Dams

The intermediate or semi-rigid buttress dams are in between the rigid and articulated buttress dams. Thus these dams eliminate the undesirable rigidity but also do not possess high degree of articulation or flexibility.

Advantages of Buttress Dams

- A buttress dam is less massive than a comparable gravity dam. The buttress dam can be constructed even on a relatively weak foundation on which the gravity dam may not be constructed.
- The spacing of the buttresses may be adjusted to utilize zones of good foundations.

- Buttress dams can be designed to accommodate moderate amounts of foundation movement without serious damages, thus it can be built in the soil with differential settlements.
- The amount of concrete required for a buttress dam is about 1/3 to ½ of the concrete required for a gravity dam of the same height.
- Access is possible to the back of the upstream face of the dam and to the foundation between buttresses for periodic inspection and for subsequent grouting and drilling pressure relied holes, if required.
- The uplift pressure acting on a buttress dam is considerably less which leads to economy in concrete and overall stability of the dam.
- The upstream face of a buttress dam being considerably inclined a large vertical component of water pressure is developed which contributes to overall stability of the dam and saves concrete.
- The power house, switch yard etc., can be located between the buttresses thus saving some cost of construction.

Disadvantages of Buttress Dams

- Skilled labor requirements and the shuttering to concrete ratio are greater for buttress dam than for a gravity dam. This may lead to higher unit rate of concrete.
- Deterioration of the upstream surface of relatively thin reinforced-concrete deck slab or arch of a buttress dam has more serious effects on the stability of the dam.
- The number of water seals to be provided and maintained for a buttress dam is usually more than for other dams.
- The foundation stresses being higher, either sound rock foundation is necessary or spread footings, or continuous mat may be required to be provided which will increase the cost of construction.
- The saving due to reduction in the quantity of concrete will also be to some extent offset by the requirement of heavy reinforcement for relatively thin members of a buttress dam.
- The buttress dams are more susceptible to willful damage (by sabotage or enemy action) than solid gravity dams.

References

- Timber-Dams-Factsheet-87529: moorsforthefuture.org.uk, Retrieved 11, June 2020
- Gravity-dam: civilknowledges.com, Retrieved 16, March 2020
- Earthen-dam: civilseek.com, Retrieved 26, April 2020
- Uses-and-application-of-a-rockfill-dam-844691: thebalancesmb.com, Retrieved 14, June 2020
- Buttress-dams-and-its-classifications: engineeringcivil.com, Retrieved 16, January 2020

Chapter 3

Hydroelectricity Generation Techniques

The electricity produced by generators that are pushed by the movement of water is known as hydroelectricity. The method of using mechanical energy of water through man-made conduits to generate electricity is known as conduit hydroelectricity. This chapter has been carefully written to provide an easy understanding of the varied facets of hydroelectricity.

The Nagarjuna dam & hydroelectric plant, India.

Hydroelectricity is electricity produced by hydropower—that is, the energy of moving water. It is the world's leading form of renewable energy. In 2005, it accounted for over 63 percent of the total renewable energy. That same year, hydroelectricity supplied about 715,000 megawatts (or 19 percent) of the world's electricity (compared to 16 percent in 2003). Although large hydroelectric installations generate most of the world's hydroelectricity, small hydro schemes are particularly popular in China, which has over 50 percent of the world's small hydro capacity.

On the downside, hydroelectric projects can disrupt ecosystems by reducing access to salmon spawning grounds and damaging habitats of birds. These projects also lead to changes in the downstream river environment, by the scouring of river beds and loss of riverbanks. Large-scale hydroelectric dams, such as the Aswan Dam and the Three Gorges Dam, have created environmental problems both upstream and downstream.

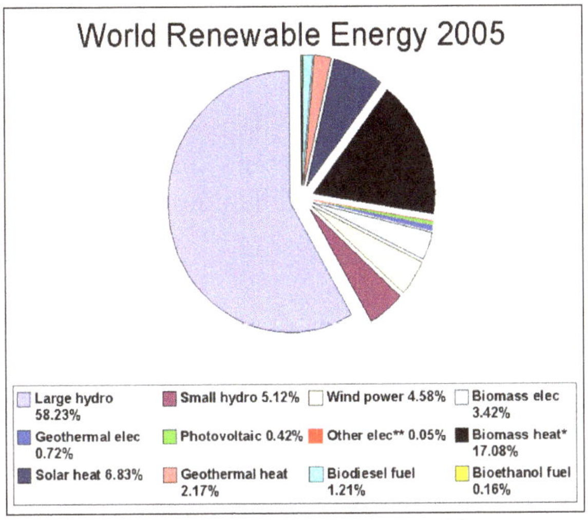

Hydroelectricity is the world's leading renewable energy source.

Electricity Generation

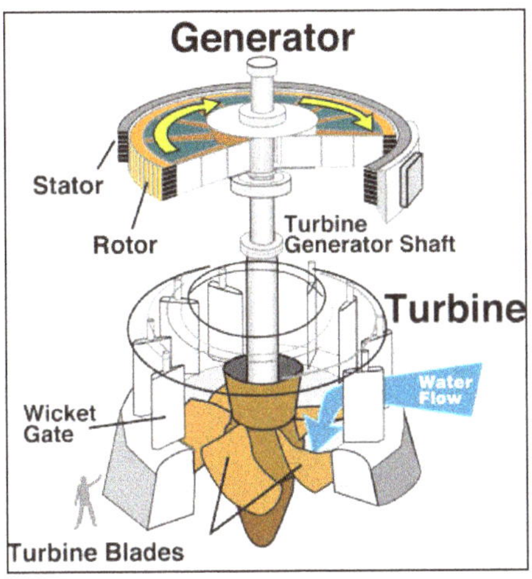

Hydraulic turbine and electrical generator.

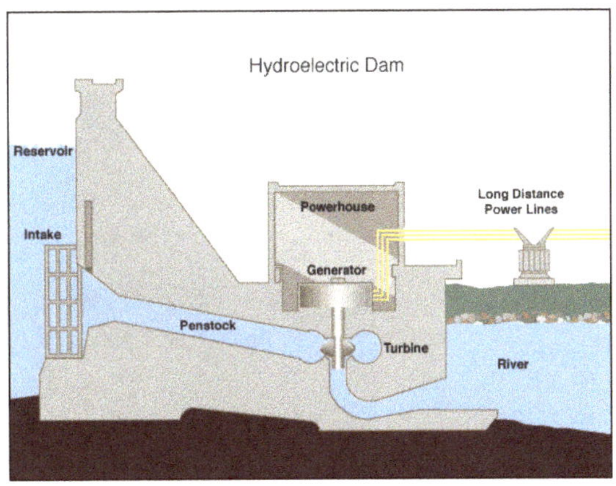

Hydroelectric dam in cross section.

Most hydroelectric power comes from the potential energy of dammed water driving a water turbine and generator. In this case, the energy extracted from the water depends on the volume and on the difference in height between the source and the water's outflow. This height difference is called the head. The amount of potential energy in water is proportional to the head. To obtain very high head, water for a hydraulic turbine may be run through a large pipe called a penstock.

Pumped storage hydroelectricity produces electricity to supply high peak demands by moving water between reservoirs at different elevations. At times of low electrical demand, excess generation capacity is used to pump water into the higher reservoir. When there is higher demand, water is released back into the lower reservoir through a turbine. Pumped storage schemes currently provide the only commercially important means of grid energy storage and improve the daily load factor of the generation system.

Less common types of hydro schemes use water's kinetic energy or undammed sources such as run-of-the-river, waterwheels, and tidal power.

Industrial Hydroelectric Plants

Many hydroelectric projects supply public electricity networks, but some are created to serve specific industrial enterprises. Dedicated hydroelectric projects are often built to provide the substantial amounts of electricity needed for aluminium electrolytic plants, for example. In the Scottish Highlands there are examples at Kinlochleven and Lochaber, constructed during the early years of the twentieth century. In Suriname, the "van Blommestein" lake, dam and power station were constructed to provide electricity for the Alcoa aluminium industry. As of 2007 the Kárahnjúkar Hydropower Project in Iceland remains controversial.

Advantages

Economics

The upper reservoir and dam of the festiniog pumped storage scheme. 360 megawatts of electricity can be generated within 60 seconds of the need arising.

The main advantage of hydroelectricity is that it is nearly independent of increases in the cost of fossil fuels such as oil, natural gas, or coal. Fuel is not required, and so it need not be imported. Hydroelectric plants tend to have longer economic lives than fuel-fired generation, with some plants now in service having been built 50 to 100 years ago. Operating labor cost is usually low since plants are automated and have few personnel on site during normal operation.

Where a dam serves multiple purposes, a hydroelectric plant may be added with relatively low construction cost, providing a useful revenue stream to offset the costs of dam operation. It has been calculated that the sale of electricity from the Three Gorges Dam will cover the construction costs after 5 to 7 years of full generation.

Related Activities

Reservoirs created by hydroelectric schemes often provide facilities for water sports, and become tourist attractions in themselves. In some countries, farming fish in the reservoirs is common. Multi-use dams installed for irrigation can support the fish farm with relatively constant water supply. Large hydro dams can control floods, which would otherwise affect people living downstream of the project. When dams create large reservoirs and eliminate rapids, boats may be used to improve transportation.

Greenhouse Gas Emissions

Since no fossil fuel is consumed, emission of carbon dioxide (a greenhouse gas) from burning fuel is eliminated. While some carbon dioxide is produced during manufacture and construction of the project, this is a tiny fraction of the operating emissions of equivalent fossil-fuel electricity generation.

Disadvantages

Environmental Damage

Recreational users must exercise extreme care when near hydroelectric dams, power plant intakes and spillways.

Hydroelectric projects can be disruptive to surrounding aquatic ecosystems. For instance, studies have shown that dams along the Atlantic and Pacific coasts of North America have reduced salmon populations by preventing access to spawning grounds upstream, even though most dams in salmon habitat have fish ladders installed. Salmon spawn are also harmed on their migration to sea when they must pass through turbines. This has led to some areas transporting smolt downstream by barge during parts of the year. Turbine and power-plant designs that are easier on aquatic life are an active area of research.

Generation of hydroelectric power changes the downstream river environment. Water exiting a turbine usually contains very little suspended sediment, which can lead to scouring of river beds and loss of riverbanks. Since turbines are often opened intermittently, rapid or even daily fluctuations in river flow are observed. For example, in the Grand Canyon, the daily cyclic flow variation caused by Glen Canyon Dam was found to be contributing to erosion of sand bars. Dissolved oxygen content of the water may change from pre-construction conditions. Water exiting from turbines is typically much colder than the pre-dam water, which can change aquatic faunal populations, including endangered species. Some hydroelectric projects also utilize canals, typically to divert a river at a shallower gradient to increase the head of the scheme. In some cases, the entire river may be diverted, leaving a dry riverbed. Examples include the Tekapo and Pukaki Rivers.

Large-scale hydroelectric dams, such as the Aswan Dam and the Three Gorges Dam, have created environmental problems both upstream and downstream.

A further concern is the impact of major schemes on birds. Since damming and redirecting the waters of the Platte River in Nebraska for agricultural and energy use, many native and migratory birds such as the Piping Plover and Sandhill Crane have become increasingly endangered.

Greenhouse Gas Emissions

The reservoirs of hydroelectric power plants in tropical regions may produce substantial amounts of methane and carbon dioxide. This is due to plant material in flooded areas decaying in an anaerobic environment, and forming methane, a very potent greenhouse gas. According to the World Commission on Dams report, where the reservoir is large compared to the generating capacity (less than 100 watts per square meter of surface area) and no clearing of the forests in the area was undertaken prior to impoundment of the reservoir, greenhouse gas emissions from the reservoir may be higher than those of a conventional oil-fired thermal generation plant.

In boreal reservoirs of Canada and Northern Europe, however, greenhouse gas emissions are typically only 2 to 8 percent of any kind of conventional thermal generation. A new class of underwater logging operation that targets drowned forests can mitigate the effect of forest decay.

Population Relocation

Another disadvantage of hydroelectric dams is the need to relocate the people living where the reservoirs are planned. In many cases, no amount of compensation can replace ancestral and cultural attachments to places that have spiritual value to the displaced population. Additionally, historically and culturally important sites can be flooded and lost. Such problems have arisen at the Three Gorges Dam project in China, the Clyde Dam in New Zealand, and the Ilısu Dam in Southeastern Turkey.

Dam Failures

Failures of large dams, while rare, are potentially serious—the Banqiao Dam failure in China resulted in the deaths of 171,000 people and left millions homeless, more than some estimates of the death toll from the Chernobyl disaster. Dams may be subject to enemy bombardment during wartime, sabotage, and terrorism. Smaller dams and micro hydro facilities are less vulnerable to these threats.

The creation of a dam in a geologically inappropriate location may cause disasters like the one of the Vajont Dam in Italy, where almost 2000 people died, in 1963.

Comparison with other Methods of Power Generation

Hydroelectricity eliminates the flue gas emissions from fossil fuel combustion, including pollutants such as sulfur dioxide, nitric oxide, carbon monoxide, dust, and mercury in the coal.

Compared to the nuclear power plant, hydroelectricity generates no nuclear waste, nor nuclear leaks. Unlike uranium, hydroelectricity is also a renewable energy source.

Compared to wind farms, hydroelectricity power plants have a more predictable load factor. If the project has a storage reservoir, it can be dispatched to generate power when needed. Hydroelectric plants can be easily regulated to follow variations in power demand.

Unlike fossil-fueled combustion turbines, construction of a hydroelectric plant requires a long lead-time for site studies, hydrological studies, and environmental impact assessment. Hydrological data up to 50 years or more is usually required to determine the best sites and operating regimes for a large hydroelectric plant. Unlike plants operated by fuel, such as fossil or nuclear energy, the number of sites that can be economically developed for hydroelectric production is limited; in many areas the most cost effective sites have already been exploited. New hydro sites tend to be far from population centers and require extensive transmission lines. Hydroelectric generation depends on rainfall in the watershed, and may be significantly reduced in years of low rainfall or snowmelt. Utilities that primarily use hydroelectric power may spend additional capital to build extra capacity to ensure sufficient power is available in low water years.

Principles of Hydropower Generation

Hydroelectric power comes from water at work, water in motion. It can be seen as a form of solar energy, as the sun powers the hydrologic cycle which gives the earth its water. In the hydrologic cycle, atmospheric water reaches the earth's surface as precipitation. Some of this water evaporates, but much of it either percolates into the soil or becomes surface runoff. Water from rain and melting snow eventually reaches ponds, lakes, reservoirs, or oceans where evaporation is constantly occurring.

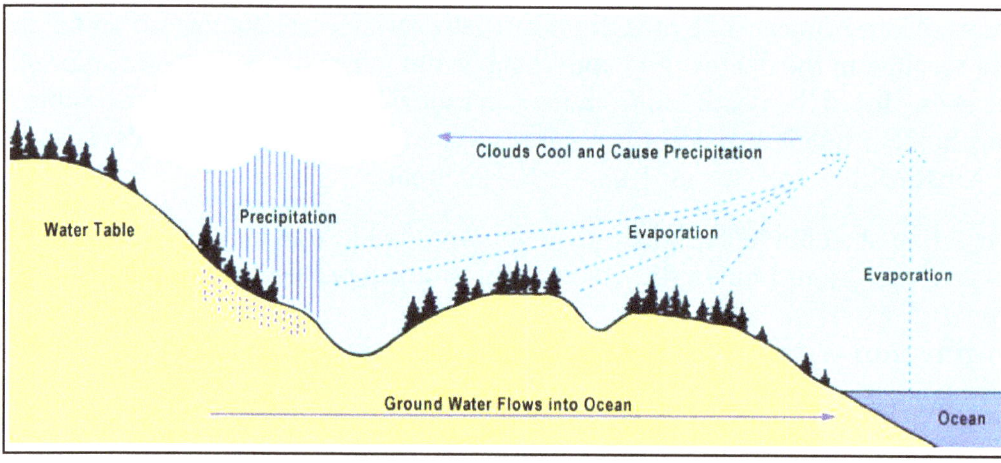

Moisture percolating into the soil may become ground water (subsurface water), some of which also enters water bodies through springs or underground streams. Ground water may move upward through soil during dry periods and may return to the atmosphere by evaporation.

Water vapor passes into the atmosphere by evaporation then circulates, condenses into clouds, and some returns to earth as precipitation. Thus, the water cycle is complete. Nature ensures that water is a renewable resource.

Generating Power

In nature, energy cannot be created or destroyed, but its form can change. In generating electricity, no new energy is created. Actually one form of energy is converted to another form.

To generate electricity, water must be in motion. This is kinetic (moving) energy. When flowing water turns blades in a turbine, the form is changed to mechanical (machine) energy. The turbine turns the generator rotor which then converts this mechanical energy into another energy form electricity. Since water is the initial source of energy, we call this hydroelectric power or hydropower for short.

At facilities called hydroelectric power plants, hydropower is generated. Some power plants are located on rivers, streams, and canals, but for a reliable water supply, dams are needed. Dams store water for later release for such purposes as irrigation, domestic and industrial use, and power generation. The reservoir acts much like a battery, storing water to be released as needed to generate power.

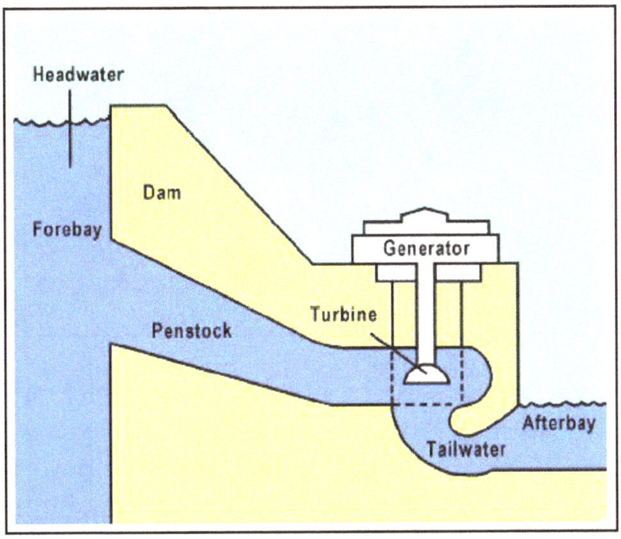

The dam creates a "head"or height from which water flows. A pipe (penstock) carries the water from the reservoir to the turbine. The fast-moving water pushes the turbine blades, something like a pinwheel in the wind. The waters force on the turbine blades turns the rotor, the moving part of the electric generator. When coils of wire on the rotor sweep past the generator's stationary coil (stator), electricity is produced.

This concept was discovered by Michael Faraday in 1831 when he found that electricity could be generated by rotating magnets within copper coils.

When the water has completed its task, it flows on unchanged to serve other needs.

Transmitting Power

Once the electricity is produced, it must be delivered to where it is needed our homes, schools, offices, factories, etc. Dams are often in remote locations and power must be transmitted over some distance to its users.

Vast networks of transmission lines and facilities are used to bring electricity to us in a form we can use. All the electricity made at a power plant comes first through transformers which raise the voltage so it can travel long distances through power lines. (Voltage is the pressure that forces an electric current through a wire.) At local substations, transformers reduce the voltage so electricity can be divided up and directed throughout an area.

Transformers on poles (or buried underground, in some neighborhoods) further reduce the electric power to the right voltage for appliances and use in the home. When electricity gets to our homes, we buy it by the kilowatt-hour, and a meter measures how much we use.

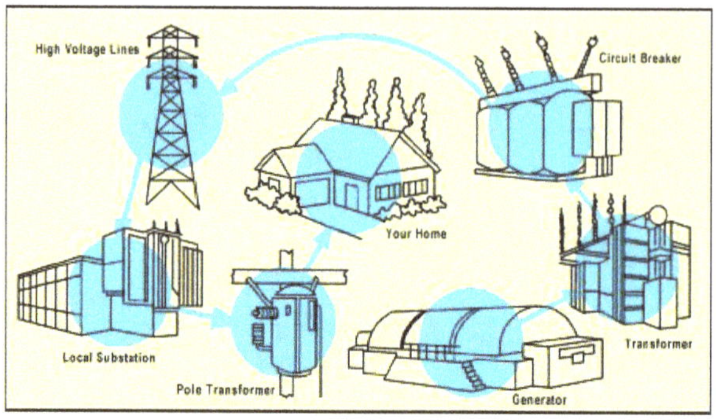

While hydroelectric power plants are one source of electricity, other sources include power plants that burn fossil fuels or split atoms to create steam which in turn is used to generate power. Gas turbine, solar, geothermal, and wind-powered systems are other sources. All these power plants may use the same system of transmission lines and stations in an area to bring power to you. By use of this a power grid," electricity can be interchanged among several utility systems to meet varying demands. So the electricity lighting your reading lamp now may be from a hydroelectric power plant, a wind generator, a nuclear facility, or a coal, gas, or oil-fired power plant or a combination of these.

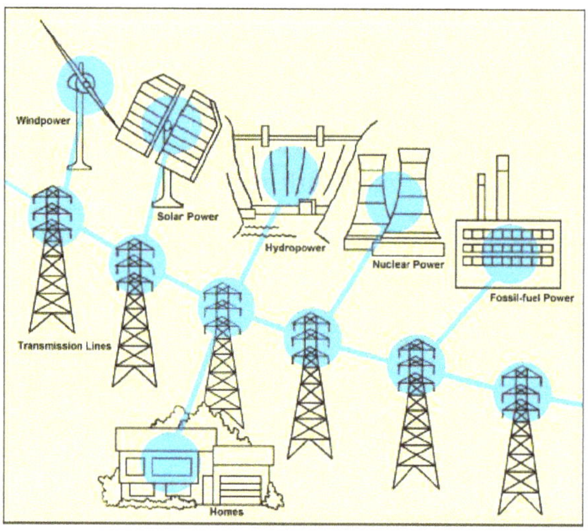

The area where you live and its energy resources are prime factors in determining what kind of power you use. For example, in Washington State hydroelectric power plants provided approximately 80 percent of the electrical power during 2002. In contrast, in Ohio during the same year, almost 87 percent of the electrical power came from coal-fired power plants due to the area's ample supply of coal.

Electrical utilities range from large systems serving broad regional areas to small power companies serving individual communities. Most electric utilities are investor-owned (private) power companies. Others are owned by towns, cities, and rural electric associations. Surplus power produced at facilities owned by the Federal Government is marketed to preference power customers (A customer given preference by law in the purchase of federally generated electrical energy which is generally an entity which is nonprofit and publicly financed.) by the Department of Energy through its power marketing administrations.

Computation of Power

Before a hydroelectric power site is developed, engineers compute how much power can be produced when the facility is complete. The actual output of energy at a dam is determined by the volume of water released (discharge) and the vertical distance the water falls (head). So, a given amount of water falling a given distance will produce a certain amount of energy. The head and the discharge at the power site and the desired rotational speed of the generator determine the type of turbine to be used.

The head produces a pressure (water pressure), and the greater the head, the greater the pressure to drive turbines. This pressure is measured in pounds of force (pounds per square inch). More head or faster flowing water means more power.

To find the theoretical horsepower (the measure of mechanical energy) from a specific site, this formula is used:

THP = (Q x H)/8.8

Where: THP = theoretical horsepower

Q = flow rate in cubic feet per second (cfs)

H = head in feet

8.8 = a constant

A more complicated formula is used to refine the calculations of this available power. The formula takes into account losses in the amount of head due to friction in the penstock and other variations due to the efficiency levels of mechanical devices used to harness the power.

To find how much electrical power we can expect, we must convert the mechanical measure (horsepower) into electrical terms (watts). One horsepower is equal to 746 watts (U.S. measure).

Turbines

While there are only two basic types of turbines (impulse and reaction), there are many variations. The specific type of turbine to be used in a power plant is not selected until all operational studies and cost estimates are complete. The turbine selected depends largely on the site conditions.

A reaction turbine is a horizontal or vertical wheel that operates with the wheel completely submerged, a feature which reduces turbulence. In theory, the reaction turbine works like a rotating lawn sprinkler where water at a central point is under pressure and escapes from the ends of the blades, causing rotation. Reaction turbines are the type most widely used.

An impulse turbine is a horizontal or vertical wheel that uses the kinetic energy of water striking its buckets or blades to cause rotation. The wheel is covered by a housing and the buckets or blades are shaped so they turn the flow of water about 170 degrees inside the housing. After turning the blades or buckets, the water falls to the bottom of the wheel housing and flows out.

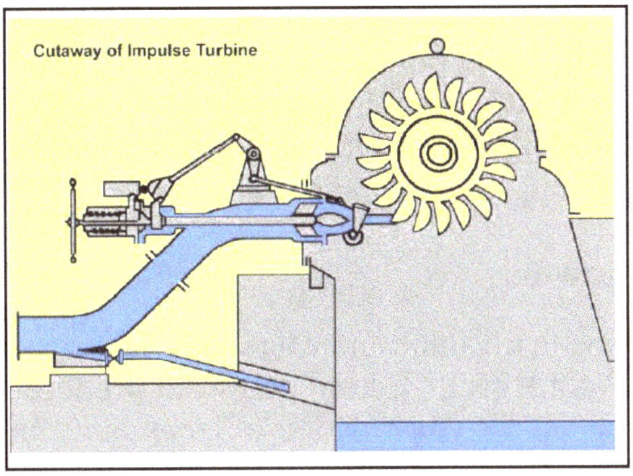

Modern Concepts and Future Role

Hydropower does not discharge pollutants into the environment; however, it is not free from adverse environmental effects. Considerable efforts have been made to reduce environmental problems associated with hydropower operations, such as providing safe fish passage and improved water quality in the past decade at both Federal facilities and non-Federal facilities licensed by the Federal Energy Regulatory Commission.

Efforts to ensure the safety of dams and the use of newly available computer technologies to optimize operations have provided additional opportunities to improve the environment. Yet, many unanswered questions remain about how best to maintain the economic viability of hydropower in the face of increased demands to protect fish and other environmental resources.

Existing hydropower concepts and approaches include:

- Uprating existing power plants
- Developing small plants (low-head hydropower)
- Peaking with hydropower
- Pumped storage
- Tying hydropower to other forms of energy

Uprating

The uprating of existing hydroelectric generator and turbine units at power plants is one of the most immediate, cost-effective, and environmentally acceptable means of developing additional electric power. Since 1978, Reclamation has pursued an aggressive uprating program which has added more than 1,600,000 kW to Reclamation's capacity at an average cost of $69 per kilowatt. This compares to an average cost for providing new peaking capacity through oil-fired generators of more than $400 per kilowatt. Reclamation's uprating program has essentially provided the equivalent of another major hydroelectric facility of the approximate magnitude of Hoover Dam and Power plant at a fraction of the cost and impact on the environment when compared to any other means of providing new generation capacity.

Low-head Hydropower

A low-head dam is one with a water drop of less than 65 feet and a generating capacity less than 15,000 kW. Large, high-head dams can produce more power at lower costs than low-head dams, but construction of large dams may be limited by lack of suitable sites, by environmental considerations, or by economic conditions. In contrast, there are many existing small dams and drops in elevation along canals where small generating plants could be installed. New low-head dams could be built to increase output as well. The key to the usefulness of such units is their ability to generate power near where it is needed, reducing the power inevitably lost during transmission.

Peaking with Hydropower

Demands for power vary greatly during the day and night. These demands vary considerably from season to season, as well. For example, the highest peaks are usually found during summer daylight hours when air conditioners are running.

Nuclear and fossil fuel plants are not efficient for producing power for the short periods of increased demand during peak periods. Their operational requirements and their long startup times make them more efficient for meeting base load needs.

Since hydroelectric generators can be started or stopped almost instantly, hydropower is more responsive than most other energy sources for meeting peak demands. Water can be stored overnight in a reservoir until needed during the day, and then released through turbines to generate power to help supply the peak load demand. This mixing of power sources offers a utility company the flexibility to operate steam plants most efficiently as base plants while meeting peak needs with the help of hydropower. This technique can help ensure reliable supplies and may help eliminate brownouts and blackouts caused by partial or total power failures.

Today, many of Reclamation=s 58 power plants are used to meet peak electrical energy demands, rather than operating around the clock to meet the total daily demand. Increasing use of other energy-producing power plants in the future will not make hydroelectric power plants obsolete or unnecessary. On the contrary, hydropower can be even more important. While nuclear or fossil-fuel power plants can provide base loads, hydroelectric power plants can deal more economically with varying peak load demands. This is a job they are well suited for.

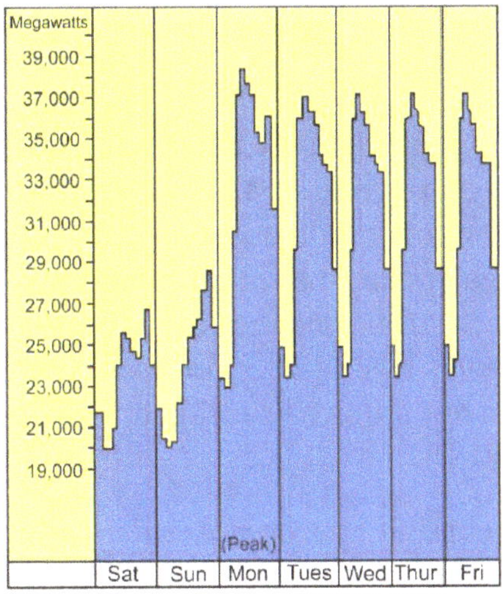

Typical Weekly Load Curve of a Large Electric Utility

Pumped Storage

Like peaking, pumped storage is a method of keeping water in reserve for peak period power demands. Pumped storage is water pumped to a storage pool above the power plant at a time when customer demand for energy is low, such as during the middle of the night. The water is then allowed to flow back through the turbine-generators at times when demand is high and a heavy load is place on the system.

The reservoir acts much like a battery, storing power in the form of water when demands are low and producing maximum power during daily and seasonal peak periods. An advantage of pumped storage is that hydroelectric generating units are able to start up quickly and make rapid adjustments in output. They operate efficiently when used for one hour or several hours.

Because pumped storage reservoirs are relatively small, construction costs are generally low compared with conventional hydropower facilities.

Conduit Hydroelectricity

Hydroelectric power generation is classified as "conduit hydropower" if the water flowing through the generating equipment has a primary use other than for power generation. For example, any power generation equipment installed on a pipeline or similar man-made water conveyance system that is operated for the distribution of water for agricultural, municipal or industrial consumption and not primarily for the generation of electricity can be classified as a conduit hydropower installation. This classification is significant because it does not require licensing or exemption by the Federal Energy Regulatory Commission (FERC).

Numerous applications for conduit hydropower exist within water systems, including pressure-reducing sites, reservoirs, treatment plants, industrial processes and irrigation. The main benefit of conduit hydropower experienced by water system operators is reduced operational costs. The power generated by a conduit hydroelectric installation can be sold to the power utility at a wholesale rate ranging anywhere from $0.04/kWh up to $0.20/kWh, depending on the prevailing cost of power at the location. If power is used at a site where electricity is consumed by a pump or other equipment, it can be "net metered." This can reduce both demand charges and the monthly electrical bill, which lowers overall operating costs. Conduit hydropower is a continuous source of energy that typically follows the same diurnal curve as power demand thereby helping to level out the power grid supply and demand. Power utilities generally prefer this type of generation over more intermittent forms of generation such as solar or wind.

Making a Site Suitable

To generate power and offset operating costs, a potential site needs to have both a pressure differential and a volume flow rate. The pressure differential available for power generation is typically the difference in height from the intake to the discharge or in the case of a control valve application, it is simply the pressure differential across the valve. If the differential pressures and flow rates are known, a conduit hydroelectric firm can provide an estimate of how much energy and revenue can be generated at the site.

A suitable site for conduit hydroelectric installations are those with high-pressure differentials and high continuous flows because they will generate the most power. However, variable flow, fixed flow and even intermittent flow applications are often cost effective as well. The potential site should have nearby power lines for interconnection. Conduit projects do not need to be new installations; they can utilize existing infrastructure such as existing vaults, piping and valving to reduce installation costs.

For example, a pressure-reducing valve that has a pressure drop of 50psi and an average flow rate of 2,000gpm could generate up to 35kW of power. If this valve operated year round it would generate 275,000kWh per year. If this power was then sold at a wholesale rate of $.10/kWh, approximately $27,500 per year of revenue would be generated. If this generation was installed at a pump station or treatment plant, for example, a net metering arrangement could be set up in which power generated would directly offset power purchased at the retail rate.

Typical System Configuration

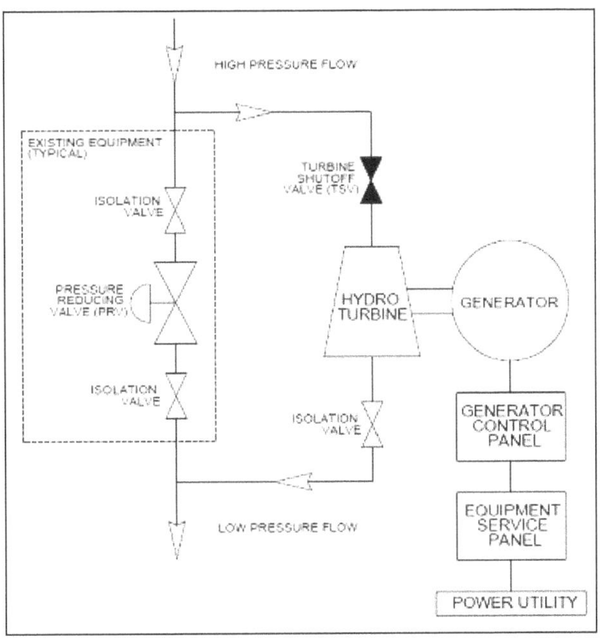

Water delivery is always the main priority of a water distribution system; therefore, conduit hydroelectric projects are configured to be a secondary priority to water delivery. When generating equipment is coupled with a custom control system, the hydro turbine functions in the same way as a control valve. Modes of operation can including pressure reducing, pressure sustaining, level control and flow control among others. The hydro turbine is designed similarly to a control valve in that it relies on inputs from the water system or SCADA to function properly. An experienced hydropower firm will take care to thoroughly understand water system operations before designing and installing a conduit hydroelectric system.

A typical hydroelectric system is plumbed parallel to a new or existing control valve. This parallel valve becomes the turbine bypass valve. When the turbine is down for maintenance or emergency, water delivery can continue uninterrupted. An automatic fail-safe valve is usually installed on the high-pressure side of the hydro turbine to operate as the turbine shutoff valve. The turbine shutoff valve allows the generating equipment to be safely shut down in the event of a power outage or emergency situation. The turbine control system automatically and seamlessly transfers the flow of water from the hydro turbine to the bypass control valve and back upon startup and shutdown.

Head and Flow

Head Hydropower all comes down to head and flow. The amount of power, and therefore energy that you can generate is proportional to the head and the flow.

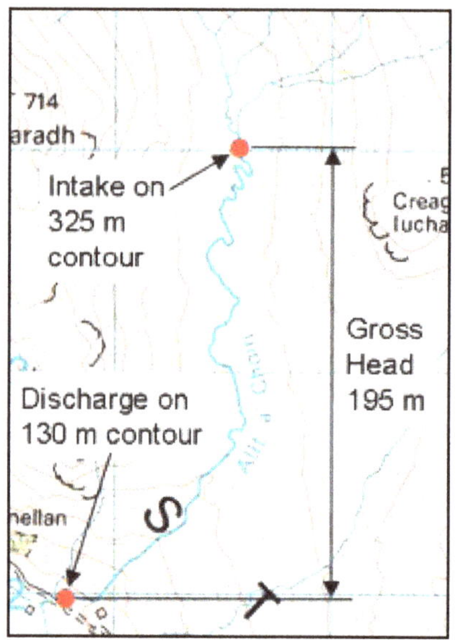

Diagram of measuring head at high head hydropower site.

Head is the change in water levels between the hydro intake and the hydro discharge point. It is a vertical height measured in metres. The two diagrams below show how the head would be measured on a typical 'low head' and a typical 'high head' site. The more head you have the higher the water pressure across the hydro turbine and the more power it will generate. Higher heads are not only better because they generate more power, but also because the higher water pressure means you can force a higher flow rate through a smaller turbine, and because turbine cost is closely related to physical size, higher-head turbines often cost less than their low-head cousins even though they might generate the same power.

Higher head also means a faster rotating turbine and generator, which means lower torque.

The cost of drive train is closely related to how much torque it has to transmit, so higher heads = less torque = less cost.

Diagram of measuring head at a low head hydropower site.

Of course you only have what you have, so if your site only has 2 ½ metres of head you won't be able to increase this significantly. However, even small increases in head can make a difference.

Sometimes it is possible to clear silt or re-grade a tailrace or discharge channel to lower the downstream water levels slightly which increases the overall head at the site. Or it may be possible to raise the water level on the upstream side by raising weir crests or sluices, though this must be done carefully to avoid increasing flood risk, and sometimes requires the construction of new spillways or installation of fail-safe tilting weirs to ensure that flood risk isn't increased during extreme flood events.

Generally speaking the cost of even small increases in head at low-head sites is repaid hundreds of times over from increased energy production for the next few decades, so is always worth the effort.

Flow

The flow rate and how it varies over a year is the next and equally important parameter.

The simplest way to characterise the flow in a watercourse is to work out what the *long-term annual mean flow* is. This is important because it is the overall average flow in a watercourse that is important; it doesn't matter if it is a raging torrent after heavy rains because in the big scheme of things we only have really heavy rains for a few days or weeks a year, and for the other 50 weeks when it is lightly raining, drizzling or bright summer sunshine you would still want you hydro system to be working and generating energy.

The fundamental piece of information that characterises the flow in a watercourse is the *flow duration curve.* Although simple once you understand them, they are quite complicated to newcomers. figure below shows the long-term flow duration curve of a small river in Somerset.

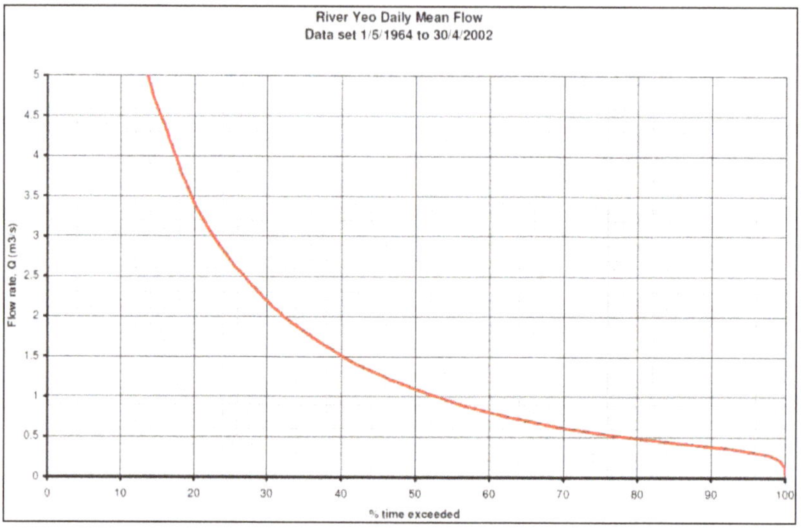

Figure: Long-term flow duration curve for the River Yeo in Somerset.

The y-axis is the flow rate in m^3/s (metres-cubed per second), or sometime in litres per second for smaller watercourses. When the flow duration curve is constructed all of the flow rate data is sorted into descending order, then the highest flow rates are plotted on the left of the curve, then progressively lower flow rates to the right until the very lowest flow is plotted at the extreme left-hand end. The x-axis is the 'percentage exceedence. This is normally the difficult part to understand. For a given percentage exceedence it shows the flow rate equalled *or exceeded* for that percentage of time. For example, if you look at the 50% percentage exceedence on figure above and read off the flow rate at that point you will see that it is 1.1 m^3/s. This doesn't mean that the flow rate in the watercourse is 1.1 m^3/s for 50% of the year, it means that the flow rate is 1.1 m^3/s *or more* for 50% of the year. This is more important because it is clear from the shape of the curve that apart from the instant that the line crosses the 50% mark it is always more than 1.1 m^3/s.

The x-axis is always plotted as a percentage exceedence from 0 to 100%. This is so that any data set spanning any interval can be plotted. The data set may span 10 days or more likely several decades of data from a local river gauging station. Generally speaking flow duration curves present long-term annual data, so the flow rates read off them are the annual flow characteristics. The percentage exceedences are often called 'Q values', so Q_{95} is the flow rate exceeded for 95% of a year and Q_{10} the flow exceeded for 10% of the year. The data set can also be analysed using a spreadsheet to work out the average (arithmetic mean) of all of the flows, and this is called Q_{mean}. In the UK the Q_{mean} is normally somewhere between Q_{25} and Q_{30}, and in the case of figure above is $Q_{26.5}$.

Once you understand how a flow duration curve is constructed it can tell you a great deal about the flow characteristics of the water course. Firstly you can work out the Q_{mean}, which is the average flow, and often if you are negotiating with the environmental regulator you will be discussing the Q_{95} flow, as this is often used as the representative 'low flow' that must always flow in any depleted stretches of river while the main volume of water passes through the hydro turbine.

By comparing different Q values you can see how 'flashy' a watercourse is, or whether it has a high base flow. The flow duration curve in figure has a Q_{mean} of 2.54 m³/s and a Q_{95} of 0.32 m³/s, so the Q_{95} is 8% of the Q_{mean}. This is typical for a 'flashy' river that rises and falls quickly in response to rain because the rain runs straight of the land and into the river without getting stored in bogs or porous rocks before flowing into the river sometime later.

If you make the same comparison using figure, you will see that the Q_{mean} is 11.30 m³/s and the Q_{95}5.83 m³/s. In this case Q_{95} is 52% of the Q_{mean}. This would be a 'high base-flow' river, typical of the rivers that flow in Hampshire with chalk catchments where the chalk stores the rain in its porous structure and then releases it to the rivers slowly and over a long period of time, rather like a sponge. It is also interesting to note that Q_{mean} is at Q_{41}; much lower on the flow duration curve than a flashy river like in figure above.

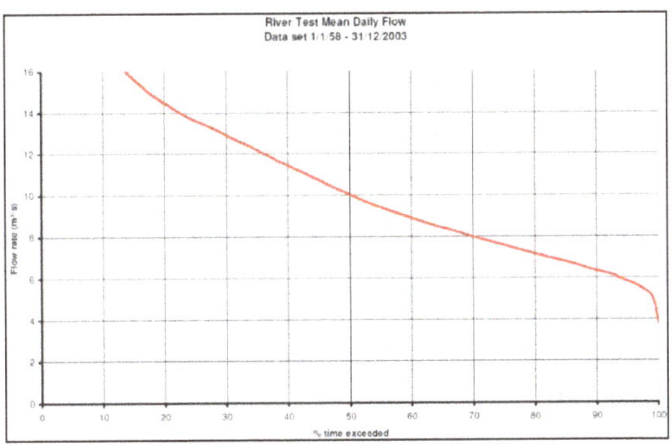

Long-term flow duration curve for the River Test in Hampshire.

In both flow duration curves you can see that the red line is heading steeply upwards at the left-hand end of the graph. This is the 'extreme flow' region, which is not much use for energy generation because such extreme flows occur for a relatively small proportion of the year, but they are very important when designing a hydropower systems to ensure that the hydro structures and turbine house don't get flooded, or even worse, washed away during the next major storm.

Minimum Head and Flow Required

For a commercially viable site it would normally need to be at least 25 kW maximum power output. For a low-head micro hydropower system you would need at least 2 metres of gross head and an average flow rate of 2.07 m³/s. To put this in context this would be a small river that was approximately 7 metres wide and around 1 metre deep in the middle.

For a site with 25 metres head a much lower average flow rate of 166 litters/second would be needed. This would be a large stream of 2 – 3 metres width and around 400 mm deep in the middle.

It is technically possible to develop smaller hydropower sites with lower power outputs, but the economics start to get challenging. This is particularly true for low-head sites; when the head drops to 1.5 metres it isn't normally possible to get any kind of return on investment, though the site could still be technically developed using Archimedean screws or modern waterwheels.

The table below shows the average flow rates needed for a range of heads from 2 metres to 100 metres for system with 100, 50, 25, 10 and 5 kW maximum power outputs. 25 kW would normally be considered the minimum for a commercial project, though a 10 kW system can still produce an acceptable return if the civil engineering works are simple (hence don't cost much). 5 kW systems are not normally viable, but the figures are shown for interest and may be useful for sites that can generate value from non-tangible benefits such as attracting visitors or positive publicity.

	Maximum Power Output (kW)				
	5	10	25	50	100
Head (m)	Flow required (m^3/sec)				
2	0.340	0.680	1.699	3.398	6.796
5	0.136	0.272	0.680	1.359	2.718
10	0.068	0.136	0.340	0.680	1.359
50	0.014	0.027	0.070	0.136	0.272
100	0.006	0.014	0.034	0.068	0.136

Minimum flow rates required for a range of (gross) heads.

Hydropower Potential

For any stretch of a watercourse, characterized by a difference in level of H meters, conveying a discharge of Q (m^3 /sec), the theoretical (potential) power,

$$N_p = \gamma QH = 1000QH \left(\text{kgm/ sec}\right)$$

$$N_p = \frac{1000}{75} QH = 13.3QH = \left(HP\right)$$

$$N_p = 13.3 \times 0.736QH = 9.8QH \left(kW\right)$$

If the rate of flow changes along a stretch, the mean value of the discharges pertaining to the two terminal sections of the stretch is to be substituted in the equation,

$$Q=(Q_1+Q_2)/2,$$

The theoretical power resources of any river or river system are given by the total of the values computed for the individual stretches,

$$N_p=\sum\frac{1000QH}{75}\times 0.736=9.8\sum QH(kW)$$

Potential water power resources can be characterized by different values according to the discharge taken as basis of computation. The conventional discharges are,

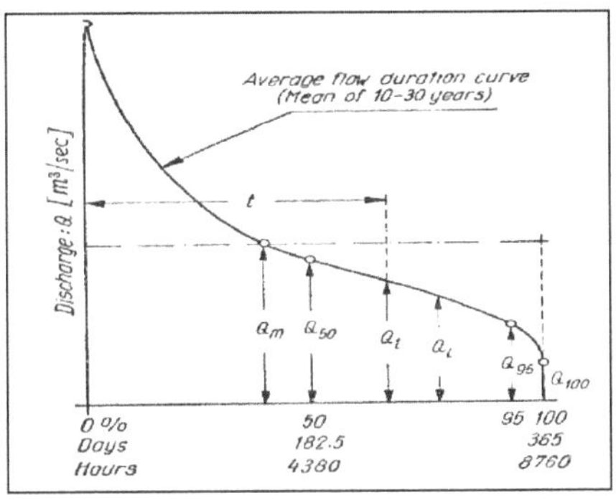

Figure: Discharges used for characterising potential water power resources.

1. Minimum potential power, or theoretical capacity of 100%, is the term for the value computed from the minimum flow observed. N_{p100}.

2. Small potential power. The theoretical capacity of 95% can be derived from the discharge of 95% duration as indicated by the average flow duration curve. N_{p95}.

3. Median or average potential power. The theoretical capacity of 50% can be computed from the discharge of 50% duration as represented by the average flow duration curve. N_{p50}.

4. Mean potential power. The value of theoretical mean capacity can be ascertained by taking into account the average of mean flow. The average of mean flow is understood as the arithmetic mean of annual mean discharges for a period of 10 to 30 years. The annual mean discharge is the value that equalizes the area of the annual flow duration curve.

Economic Significance of Potential Water Resources of a Site

This is influenced by a great number of factors than hydraulic, such as geographical, geological and topographical conditions, energy demand, etc. Ignoring these and comparing relative values of power potential as reflected by hydraulic conditions only, the following four aspects are to be taken into consideration:

a) The absolute quantity of theoretical water power resources.

b) The relative share of discharge in the power. Among the hydraulic possibilities representing equal magnitudes, the more advantageous are those where the power in question originates from a smaller flow and a higher head. It is advantageous of highland developments over power stations situated in hilly regions or lowland areas.

c) The relative annual fluctuation of available potential power.

This can be characterized by the ratio of the values Np_{50} to the values Np95 (or Np_{100}).

$$\alpha = \frac{N_{p50}}{N_{p95}} \text{ or, } \quad \alpha_1 \frac{N_{p50}}{N_{p100}}$$

A smaller ratio reflects a more favorable hydraulic possibility.

d) The over year or multi-annual variation of potential power.

This can be characterized either by a simple diagram showing the annual potential power against time, or by a summation curve of annual values.

Power resources can be characterized even by annual values of potential energy in a river, by the quantities of work,

E_{100} , E_{95}, E_{50} and E_m

All expressed in kilowatt-hours. These values can be computed as areas of the lower parts of the potential power-duration curves. If the head is assumed to be constant, independent of discharge, the computation can be based on the discharge-duration curve. Using the curve,

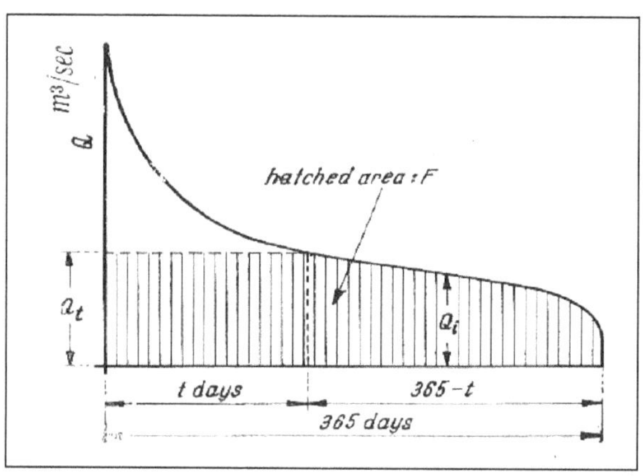

$$N_p = 9.8H\alpha Q_1 = \alpha Q_1 (kW)$$

$$E = 24\alpha Q_1 t + 24\sum_{t}^{365} \alpha Q_1$$

$$E = 24\alpha\left(Q_1 t + \sum_{t}^{365} Q_1\right) = 24\alpha F(kWh)$$

Where,

t = the duration considered in days,

Q_t = Selected discharge,

Q_i = Daily mean of actual discharge at any time,

F = Area pertaining to Q_t (shaded area).

The upper limit of potential energy inherent in the river section is obtained by,

$$E_{mx} = (24 \times 365) N_m = 8760 N_m (kWh)$$

Where N_m is the annual mean power.

The overall coefficient is about 0.75 or 0.80. The equation recommended is,

$$N_{mnet} = (7.4 - 8.0)\sum_{t}^{365} Q_m H$$

Q_m = the arithmetic mean discharge. Net river energy potential,

$$E_{max} = 8760 \sum N_{m\,net}$$

For characterizing the gross potential power of a river basin, the following data should be used:

a) Total annual discharge volume V (m^3),

b) Medium height of the watershed area H lying upstream above sea level (m),

c) Area A of this basin (km^2).

Since 1 m^3 water weighs 1 ton, the product VH yields the annual gross power pertaining to the selected site in meter-tons,

$$E(kWh) = \frac{VH}{367}$$

$$e(kWh/km^2) = \frac{VH}{367A}$$

If the e values are determined along the river basin at several stream sites, then the lines connecting the points of equal e, isopotential lines can be drawn.

Flow-Duration Curve

If the flows for any unit time are arranged in descending order of time (without regard to chronological sequence), the percentage of time for which any magnitude is equaled or exceeded may be computed. The resulting array is called a flow-duration curve.

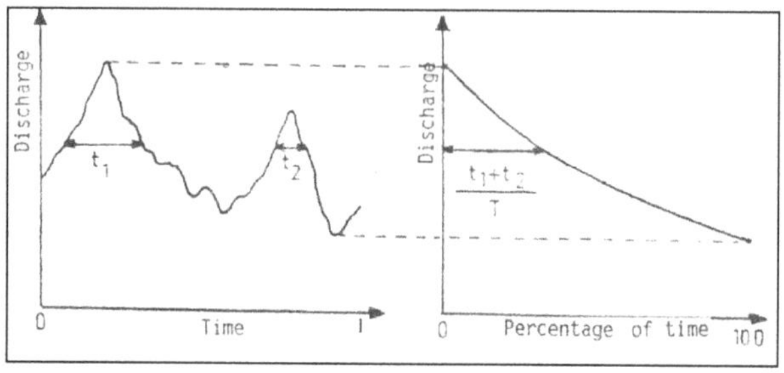

Figure: Determination of flow-duration curve.

Such curves are useful in determining the relative variability of flow between two points in a river basin or between two basins. For example, if a stream is highly regulated, the curve will approach a horizontal line. The dependable flow is that corresponding to 100 percent of time. The relative variability of two flow records may be compared by converting the discharge scale in terms of a ratio to the mean. Any sub area under the curve represents the volume of annual runoff.

Flow-duration curves have been used to approximate the amount of storage needed to increase the dependable flow. For example, the horizontal line AB in below figure may represent a new dependable flow, and the required storage needed to obtain this flow is indicated by area ABC.

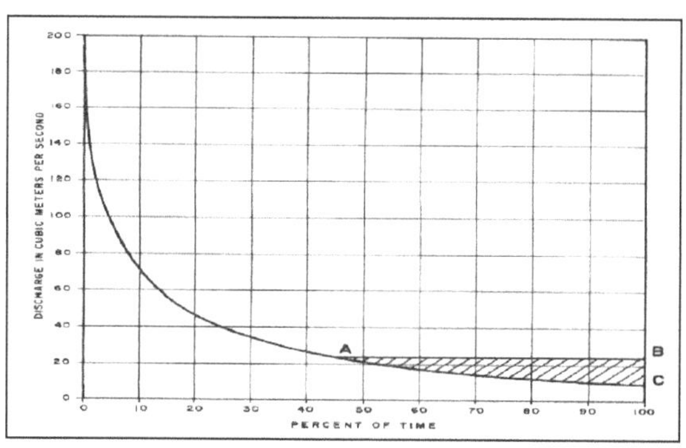

Figure: Duration curve of monthly discharges.

Power production values may be approximated from the duration curve by converting the discharge scale to kilowatts by multiplying by a selected head, efficiency and conversion factors. If the time scale is converted to hours in a year, a unit of are represents kilowatt-hours.

The flow-duration curve is particularly useful in combination with a sediment rating curve (river discharge versus the transported sediment load usually expressed in tons per day), to compute total sediment load to be expected in an average year.

Flow Mass Curve

Total flow volume from a certain initial time t = 0 up to time t can be computed as,

$$H = \int_0^t Q dt$$

In practice, the total volume is computed as,

$$H = \sum Q_1 t_1$$

Q_i = the average discharge in time interval (month, year) Δt_i.

Flow mass curve is a plot of the cumulative runoff from the hydrograph against time. The time scale is the same as for the hydrograph and may be in days, months or years. The volume ordinate may be in m³ -days, m³ -months, m³ -years, etc. The slope of the mass curve is the derivative of the volume with respect to time or the rate of discharge.

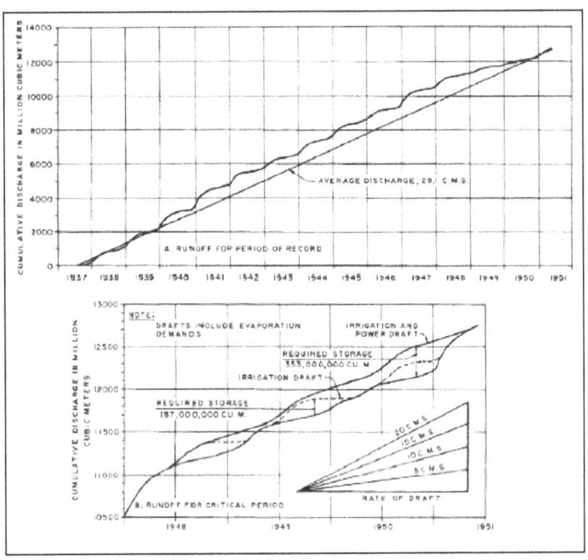

The mass curve usually has a wavy configuration in which the steeper segments represent high flow periods and flatter segments represent low flows. Uniform rates of withdrawal (draft) may be represented as tangent lines drawn from high points to intersect the curve at the next wave. The vertical distance between the draft line and the basic curve represents the cumulative difference between regulated outflow and natural inflow, or the required storage. If the draft line does not intersect the mass curve at the end of a year, it means that the reservoir does not refill with that rate of draft and regulation at the proposed draft rate will extend over two years or more. A typical mass curve is shown in the above figure.

In estimating storage requirements from the mass curve, it is not necessary to assume a constant rate of regulated flow. For example, if the draft rate to meet a demand for irrigation, water supply, or power varies from month to month, the draft line may be a curved or irregular line and the maximum draft may not occur at the low point in the mass curve.

An allowance for evaporation should be applied to the mass curve analysis. If the water area does not change significantly during the annual cycle of use, an average correction for each calendar month can be subtracted from the inflow or added to the draft rates.

The ordinates of the flow mass curve increase continuously in time. The sum of the differences between the inflow and the yield (average flow) are drawn;

$$H_o = \sum (Q_i - Q_{ave}) \Delta t_1$$

Reservoir capacity is then vertical distance between the highest and lowest points of the curve.

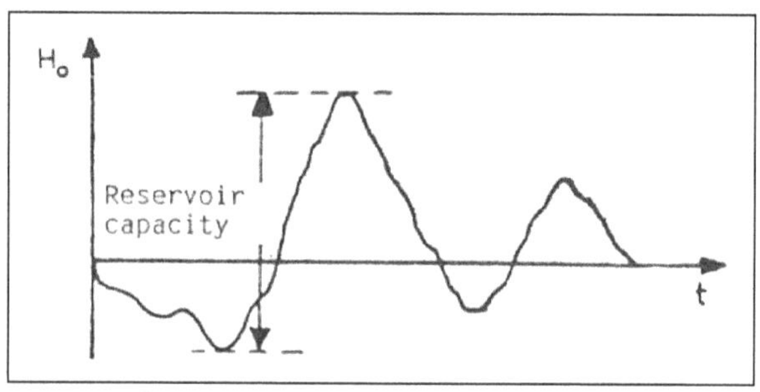

Figure: Flow mass curve derived using the differences of the discharges from the yield.

Storage-Draft Curve

The results of a mass-curve analysis can be plotted as a storage-draft curve. His curve gives the storage needed to sustain various draft rates. Examples of storage-draft curves are shown in the below figures. Both irrigation requirements and combined irrigation and power requirements are illustrated. These curves were computed from the mass curve.

If storage unlimited, the storage-draft curve will approach the available mean flow as asymptote. It is rarely possible to develop mean annual flow of a river basin. For most projects, some spillage will occur in years of runoff. To impound all flood flows will require an extensively large reservoir. Such a reservoir may not fill in many years, and probably could not be justified economically. The selected rate of regulated flow to be developed will depend on:

1. The demand of water users,
2. The available runoff,
3. The physical limits of the storage capacity,
4. The overall economies of the project.

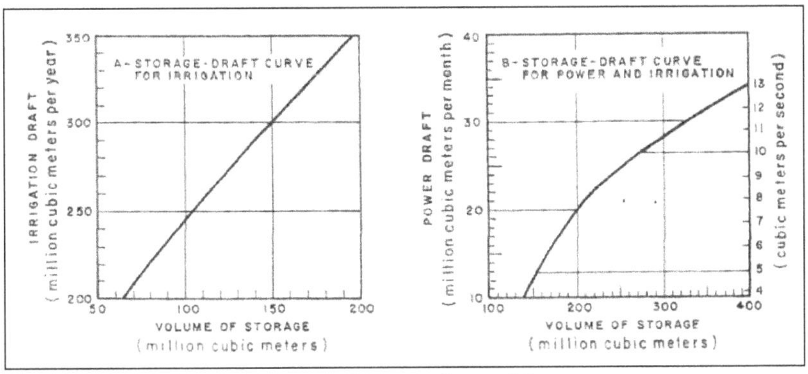

Figure: Storage-draft curves for multipurpose uses.

Selection of Design Flow

The hydrologic analyses, combined with economic analyses of costs and benefits for different heights of dam and reservoir capacity will lead to the selection of the reservoir capacity and the corresponding dependable flow that can be justified. The selected design flow may not necessarily be available 100 percent of the time. The propose water use may permit deficiencies at intervals, for example, a 15 percent shortage once in 10 years. Irrigation water supplies may permit greater deficiencies than those for urban and industrial use. Hydroelectric power plants, connected to large systems, may tolerate substantial water supply deficiencies.

Final Storage Selection

a) Evaporation Losses

Detailed evaluation of evaporation losses should be postponed until final operation and routing studies, when the actual variation in water area can be considered as well as the seasonal variation in evaporation.

Basic data on water surface evaporation may be obtained from records of pan evaporation. Such records overestimate lake evaporation and must be reduced by a pan coefficient which varies from 0.60 to 0.80 depending on the climate. The collection of evaporation records at a project site should be initiated in the planning stage. Evaporation corrections should be made on a monthly basis using actual past precipitation records at the project site if possible.

b) Power

Selection of an average flow alone will not permit determination of the benefits from a water resources development project without more detailed studies. Such studies require routing through the reservoir the entire record of flow (corrected for evaporation losses), on a month by month basis, using assumed patterns of use, outlet capacities and, in the case of power, turbine and generator capacities and efficiencies. The reservoir would normally be considered to be full at the start of the operation study, or at least full to normal pool.

For power benefits, the energy output will vary in accordance with the inflow, outflow, and change in storage and corresponding head, tail water elevation, turbine capacity and plant efficiency. If the plant is a part of a system, the output may be subject to varying demands of the system load curve and whether the plant is to be used as a base load plant or a peaking plant. The routing study will indicate the necessary modifications to the head, storage, and even height of the dam to obtain maximum benefit.

c) Irrigation

Operation studies for irrigation use should be made using seasonal crop demands and selected outlet capacities. Short-term demands may indicate that the storage needed war greater than that required for uniform regulated flow. The proposed annual water use may be greater than that available 100 percent of the time, with the understanding that deficiencies can be tolerated in some years.

d) Water Supply

Operation studies for projects providing urban water supplies will be similar to those for irrigation projects in that there may be variations in the seasonal demand, especially where more than one source is available, or where there can be transfers to other regulation reservoirs. However, the degree of dependability of flow must be higher for urban water supply than for irrigation projects.

e) Flood Control

The storage allocated for flood control in single purpose or multipurpose projects is usually based on a definite design flood the control of which is needed for downstream protection. The required storage capacity is based on routing of the design flood inflow coincident with releases not to exceed downstream channel capacities.

Total Storage Requirement

The total usable storage needed for multipurpose for multipurpose projects require more complex routing studies and numerous trials to obtain the most economic allocations.

In addition to the variable requirement for storage for downstream uses, the total storage may be increased for the following reasons:

- Minimum head on power installations.
- Allowance for the storage of sediments without loss of usable storage.
- Minimum area for recreation use, including seasonal requirements.

Example: Monthly flow volumes feeding a reservoir are given in the table. Determine the storage capacity required to supply the mean annual flow volume yield.

Solution: Cumulative volumes are calculated and given in the table,

Month	Volume ($10^6 m^3$)	Cumulative Volume ($10^6 m^3$)
1	296	296
2	386	682
3	504	1186
4	714	1900
5	810	2710
6	1154	3864
7	746	4610
8	1158	5768
9	348	6116
10	150	6266
11	223	6489
12	182	6671

Total volume of flow feeding the reservoir is 6671×106 m^3 . Annual mean discharge can be calculated as,

$$Q = \frac{6671 \times 10^6}{365 \times 86400} \cong 212\ m^3/s$$

The reservoir storage capacity required to obtain 212 m^3 /s yield throughout the year is found by drawing tangents parallel to the average draft line from peak points. The vertical distance is 1800×106 m^3 is the required capacity of the reservoir.

The reservoir capacity to supply the annual mean discharge can be found out by using the sum of differences method as in table,

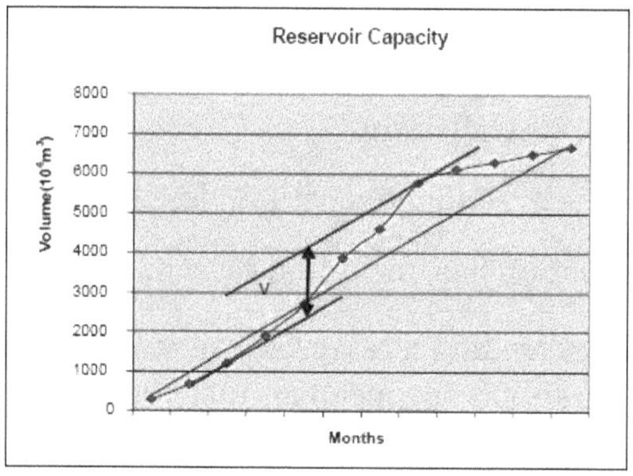

Month	Volume ($10^6 m^3$)	Flow (m^3/s)	Ho (m^3/s)	ΣHo (m^3/s)
1	296	111	-101	-101
2	386	149	-63	-165
3	504	188	-24	-188
Month	Volume (106m3)	Flow (m3/s)	Ho (m3/s)	ΣHo (m3/s)
1	296	111	-101	-101
2	386	149	-63	-165
3	504	188	-24	-188

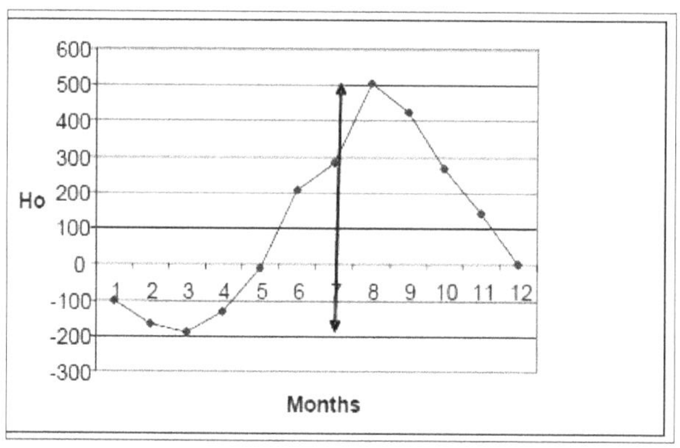

$$H_o = \sum (Q_i - Q_{ave}) \Delta t_1$$

$$\text{Reservoir Capacity} = 504 - (-188) = 693\ \text{m}^3/\text{s}$$

$$\text{Volume} = 692 \times 31 \times 86400 = 1853 \times 10^6\ \text{m}^3$$

Reservoirs

A reservoir is a manmade lake or structure used to store water. A dam reservoir has an uncontrolled inflow but a largely controlled outflow. The water available for storage is totally a function of the natural stream stream-flow that empties into it.

Reservoir Capacity

Reservoir capacity is the volume of water that can be stored in the particular reservoir. It is the normal maximum pool level behind a dam. This can be calculated by using a topographic map of the region. First, the area inside different elevation contours is measured, and then a curve of area versus elevation can be constructed.

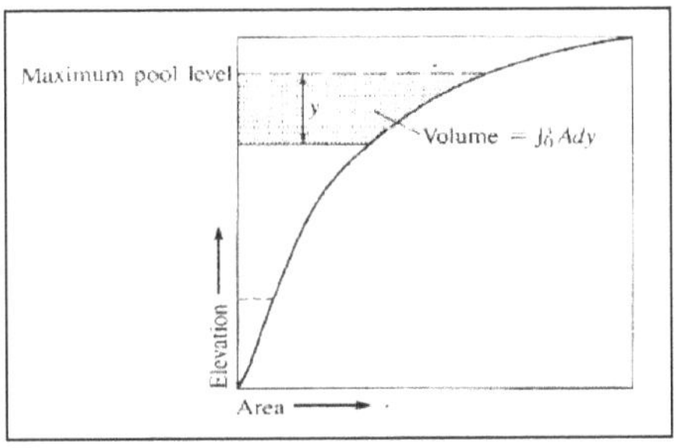

Figure: Area versus elevation for a reservoir.

At any given elevation, the increment of storage in the reservoir at that elevation will be Ady, where dy is a differential depth. Then the total storage below the maximum level to any will be given by,

$$V = \int_0^y A dy$$

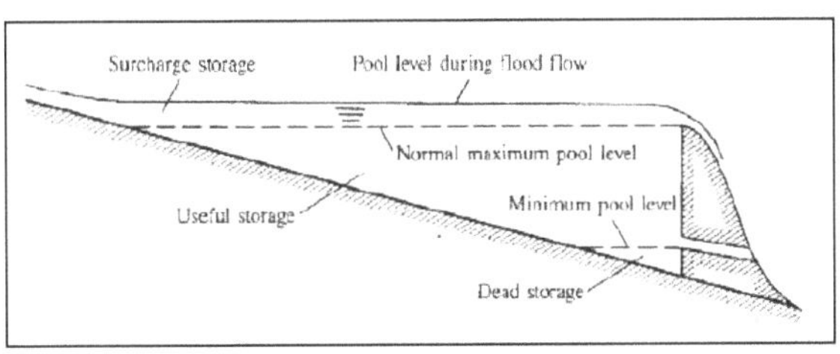

Figure· Storage relations for a reservoir with an uncontrolled spillway.

Sedimentation in Reservoirs

All streams carry sediments that originate from erosion processes in the basins that feed the streams. After a dam is constructed across the steam and a reservoir is produced, the velocity in the reservoir will be negligible so that virtually all the sediment coming into the reservoir will settle down and be trapped. Therefore, the reservoir should be designed with enough volume to hold the sediment and still operate as a water storage reservoir over the project's design life. For large projects, the design life is often considered 100 years.

Sediment carried in a stream is classified as either *bed load* or *suspended load*. The bed load consists of the coarsest fractions of the sediment (sands and gravels), and it rolls, slides, and bounces along the bottom of the stream. The finer sediments are suspended by the turbulence of the stream. When the sediment enters the lower velocity zone of the reservoir, the coarser sediments will be deposited first, and it is in this region that a delta will be formed. The finer sediments will be deposited beyond the delta at the bottom of the reservoir.

The total sediment outflow from a watershed or drainage basin measured in a specified period is the *sediment yield.* The yield is expressed in terms of tons per square kilometer per year. The engineer designing a reservoir must estimate the average sediment yield for the basin supplying the reservoir to determine at what rate the reservoir will fill with sediment.

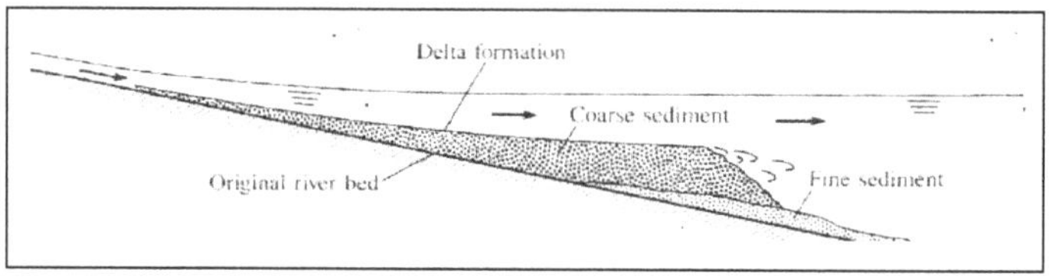

Figure· Deposition of sediment in a reservoir.

For a given reservoir volume, V, the ratio of the reservoir volume decrease due to the deposited sediment can be estimated by this empirical equation,

$$R = 23\times 10^{-6} G^{0.95}\left(\frac{A}{V}\right)^{0.8}$$

Where,

G = Sediment yield of the basin (kN/km^2/year) A = Drainage basin area (m^2)

V = Reservoir volume (m^3)

Multiplying R ratio with the design life of the reservoir, T, will yield the percentage of the dead volume in the reservoir. The dead volume can be estimated over the period of design life by,

$$V_{dead} = R\times T\times V_{reservoir}$$

EXAMPLE: The total volume of a reservoir is V = 230×10^6 m^3 with a drainage basin of A = 1200 km^2. The design life of the project is T = 100 year and the density (specific mass) of the deposit is ρ = 2.65 ton/m^3. Calculate the dead volume of the reservoir.

Solution: The sediment of the river for a river,

$$G = 1421A^{-0.229}$$
$$G = 1421\times 1200^{-0.229}$$
$$G = 280\left(m^3/km^2/\text{year}\right)(\text{volume})$$

The ratio of the reservoir volume decrease every year,

$$R = 0.000023 \times G^{0.95}\left(\frac{A}{V}\right)^{0.8}$$

$$R = 0.000023 \times 7279^{0.95} \times \left(\frac{1200 \times 10^{6}}{230 \times 10^{6}}\right)^{0.8}$$

$$R = 0.4(0/0)$$

Reservoir volume decrease due to the sediment deposit every year,

$$V_{dead} = 0.004 \times 230 \times 10^{6} = 0.92 \times 10^{6}\ m^{3}$$

For the 100 year of design period,

$$V_{dead100} = 100 \times 0.92 \times 10^{6} = 92 \times 10^{6}\ m^{3}$$

Useful storage is,

$$V_{useful} = (230 - 92) \times 10^{6} = 138 \times 10^{6}\ m^{3}$$

Wind-Generated Waves, Setup and Freeboard

Whenever wind blow over an open stretch of water, waves develop, and the mean level of the water surface may change. The latter phenomenon, called *setup* or *wind tide*, is significant only in relatively shallow reservoirs. When a dam is designed, the crest of the dam must be made higher than the maximum pool level in the reservoir to prevent overtopping of the dam as the wind-generated waves strike the face of it. The additional height given to the crest of the dam to take care of wave action, setup, and possibly settlement of the dam (if it is earth fill) is called *freeboard.*

Setup

Consider the basin of water shown in the figure. The solid line depicting the water surface is the case when no wind is blowing; the water surface is horizontal. When the wind is blowing, a shear stress acts on the water surface, and because of this, the surface will tilt, as shown by the broken line in the below figure.

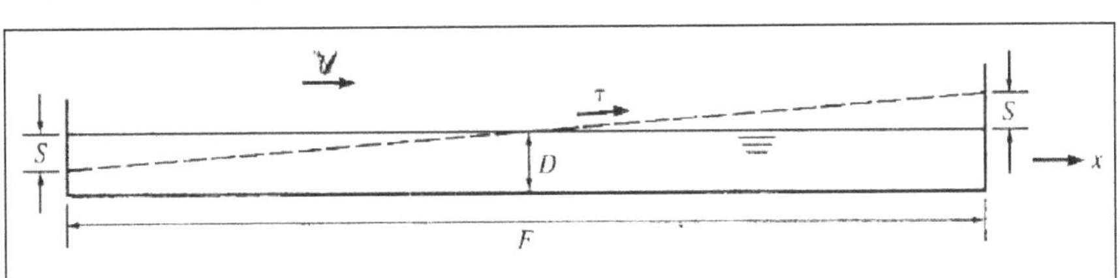

Figure· Definition sketch for setup.

The amount of setup S is,

$$S = \frac{V^2 F}{KD}$$

Where,

D = Average reservoir depth (m),

V = the wind speed measured at a height of 10 m from the surface (km/h) F = the wind fetch (km)

K = A constant ≈ 62000

S = Setup (m)

EXAMPLE: A reservoir is oval shaped with a length of 20 km and a width of 10 km. If the wind blows in a direction lengthwise to the reservoir with a velocity of 130 km/h, what will be the setup of the average water depth of the reservoir is 10 m.

Solution: The setup will be,

$$S = \frac{V^2 F}{KD}$$

$$= \frac{130^2 \times 20}{62000 \times 10} = 0.55m$$

Height of Wind Waves and the Run-Up

Allowances for wave height and the run-up of wind-generated waves are the most significant components of freeboard. The *run-up* of the waves on the upstream dam face, i.e. the maximum vertical height attained by a wave running up a dam face, is equal to H (wave height) for a typical vertical face in deep water, but can attain values over 2H for a smooth slope 1 in 2.

A *wave height*, H (m), (crest to trough) can be estimated by,

$$H = 0.34\sqrt{F} + 0.76 - 0.26\sqrt[4]{F}$$

Where,

F = the fetch length (km),

H = Wave height (m)

For large values of fetch (F>20 km), the last two terms may be neglected. With the provision for the wind speed, the equation takes the form of,

$$H = 0.32\sqrt{VF} + 0.76 - 0.26\sqrt[4]{F}$$

Where,

V = Wind velocity (km/hour)

The freeboard will be equal to set-up plus run up allowance for settlement of the embankment plus and amount of safety (usually 0.50m).

EXAMPLE: Calculate the wind set-up and wave height for a reservoir with 8 km fetch length. The average reservoir depth is 15 m. The wind velocity is V = 100 km/h. If the upstream of the dam is vertical, what will be the minimum freeboard to be given?

Solution:

The wind set-up is,

$$S = \frac{V^2F}{KD}$$

$$= \frac{100^2 \times 8}{62000 \times 15} = 0.9m$$

The wave height,

$$H = 0.32\sqrt{VF} + 0.76 - 0.26\sqrt[4]{F}$$

$$H = 0.032\sqrt{100 \times 8} + 0.76 - 0.26\sqrt[4]{8}$$

$$H = 1.19m$$

Since the upstream side of the dam is vertical, the run-up height will be taken as the height of the wave. The freeboard,

$$H_{freeboard} = 0.09 + 1.19 + 1.19 + 0.5$$

$$H = 2.97m \cong 3.00m$$

0.5 m is the safety height.

Tidal Energy Generation

Tidal energy or tidal power is a form of renewable energy obtained due to alternating sea levels. The kinetic energy from the natural rise and fall of tides is harnessed and converted into electricity. Tides are caused by the combined gravitational forces of the moon, sun, and earth. However, tides are influenced most by the moon. The moon's gravitational force is so strong that it tugs the ocean into bulge. The high and low tides create tidal currents, which are essential in generation of this kind of energy mostly prevalent in coastal areas. Tidal energy is considered a renewable energy resource because the oceans and seas will remain until the end of time and tides are highly predictable.

Tidal energy generation plants are most commonly installed along coastlines although offshore plants are increasingly becoming popular. Coastlines are preferred because they receive 2 high tides and 2 low tides every single day. To generate electricity, the disparity in water levels must be at least 5 meters.

Tidal power has great potential for future as tides can be much more accurately predicted than wind or sun and due to massive size of oceans. Though it available in plenty but harnessing energy from it is not that easy. It suffers from huge investment and limits availability of sites where it can be captured.

Conversion of Tidal Energy into Electricity

Tidal energy is converted into electricity using three main tidal technologies:

1. Tidal Turbines

Tidal turbines utilize the same technology to wind turbines. The only difference is that the blades of tidal turbines are way stronger and shorter. So, the best way to compare tidal turbines is underwater windmills. Ideally, the water currents turn the turbine. The turbine is connected to a generator through a shaft. So, when the turbine turns, the shaft also turns. The turning shaft activates a generator, which generates electricity. The initial cost of setting up this tidal stream system is quite on the higher side, not to mention the difficulty in maintenance. However, it remains a cheaper alternative and doesn't cause environmental degradation compare to other tidal technologies.

2. Tidal Barrages

Tidal Barrages are the most efficient tidal energy technologies. They resemble dams used in hydropower plants. The difference is they are a lot bigger since they are constructed across a Bay or an Estuary. For the barrage to be able to produce power, the tidal range, which is the difference between low and high tide, has to be more than 5 meters. As the tide enters the system, ocean or sea water flows via the dam into the basin. When the tides subside, the system's gates close, trapping the water in the estuary or basin. When the tides start to move out, the gates in the dam that consist of turbines, open up, and water begins to flow out hitting the turbines, which eventually turn to produce energy. Construction of tidal barrages involves high upfront capital costs, plus they have devastating effects on the local environment.

3. Tidal lagoons

This technology has a lot in common with tidal barrages. It's just that it doesn't involve a lot of initial capital outlay and it's friendly to the environment. Tidal lagoon is a power station separated from the rest of the ocean or sea. Its functionality is similar to tidal barrage since when the tide goes up, the lagoon completely fills up. When the tide subsides, the water is allowed to drain out through an opening consisting of turbines. The outward flow of water turns the turbine, which generates energy.

Advantages of Tidal Energy

By taking the time to critically analyze the advantages and disadvantages of tidal energy, you can get a clear picture of whether it's a sustainable energy resource or not. Below is an outline of the upsides and downsides to tidal energy.

1. It's environmental friendly

The fact that tidal energy technologies are installed on the coastlines and offshore makes them good for the environment since land will not be interfered with. Also, tidal energy is a clean source of energy, meaning it doesn't release any greenhouses gasses to the atmosphere.

2. It's a renewable energy

Tides harnessed to produce tidal energy result from the combined gravitational pull of the sun, moon, and earth in conjunction with the rotation of the planet around its axis. This is a natural process that occurs every single day. This means that tides will continue to occur and production of tidal energy will continue until the end of time.

3. It's highly predictable

Development of tides is a well-understood cycle. This makes it a lot easier to develop tidal energy systems with the right dimensions. Why? Because the level of power the system will be exposed to is already determined. Which explains why the capacity of the installed equipment and the entire physical size has completely no energy generation limitations, even though stream generators and tidal turbines used resemble those of wind turbines.

4. It's Cost-competitive

Tidal energy technologies once constructed have the potential to generate electricity for many years, which means they are long lasting. Although the upfront costs of setting up a tidal power plant are relatively high, the return on investment will be realized in the long run. A typical example is the La Rance tidal barrage, which is still producing electricity since 1966.

5. It Minimizes over-dependence on fossil fuels

Fossil-based sources of energy such as oil, coal and natural gas emit greenhouse gasses that lead to climate change and global warming. Tidal energy offers a green and renewable alternative to cut back on greenhouse gas emissions.

6. It Offers a sense of protection

Barrages and dams that are utilized to tap tidal energy for the generation of electricity could insulate coastal areas and ship ports from high impact and dangerous tides in the course of bad weather and storms.

7. It's quite effective than wind even at low speeds

Oceans currents have the capacity to produce more energy than air currents because ocean water is 832 times denser than air. This means ocean currents applies greater force on the turbines to generate more energy.

Disadvantages of Tidal Energy

1. High upfront capital costs

Tidal energy technologies are considerably new. Meaning the costs of infrastructure are relatively high at the moment. Experts also project that tidal energy will only start to be commercially beneficial in 2020 with enhancements of innovative technologies.

2. completely environmentally friendly

Tidal energy generation systems are thought to harbor some environmental impacts, but they have not been quantified. In addition, these tidal plants produce electricity using tidal barrages that depend on manipulation of sea levels. This means that they have the same environmental impacts as hydroelectric dams.

3. Problems of efficiency

Generation of tidal electricity wholly depends on tidal surges, which happen twice a day. This means when tides are not happening, there is no production of energy, which is why extra costs must be incurred to set up energy storage systems.

4. Tidal energy needs Long gestation period

Tidal power plants need a lot of time to be able to produce electricity efficiently. This aspect combined with cost of installation can be unsustainable. A typical example of a tidal power plant that was closed due to time and cost overruns is the UK's Severn Barrage.

5. Impact on marine life

The greatest fear among tidal energy systems developers is the impact the plants and turbines will have on the surrounding marine ecosystem. The rotation of turbines and vibrations of the tidal plant could significantly interrupt marine ecosystem and inhibit natural movement of marine life.

Tidal Current

Tidal currents, as their name suggests, are generated by tides. Tides are essentially long, slow waves created by the gravitational pull of the moon, and to a lesser degree, the sun, on the earth's surface. Since the moon is so much closer to the earth than the sun, its pull has more influence on the tides.

The moon's gravitational pull forces the ocean to bulge outwards on opposite sides of the earth, which causes a rise in the water level in places that are aligned with the moon and a decrease in water levels halfway between those two places. This rise in water level is accompanied by a horizontal movement of water called the tidal current.

Tidal currents also switch directions every time the tide transitions between high and low. Although tides and tidal currents don't have much impact in the open oceans, they can create a rapid current of up to 15.5 miles (25 kilometers) per hour when they flow in and out of narrower areas like bays, estuaries and harbors. Fast tidal currents toss sediment around and affect plant and animal life. Currents may, for example, transfer a fish's eggs from an estuary out into the open sea or carry nutrients that the fish needs from the sea into the estuary.

The strongest tidal currents occur at or around the peak of high and low tides. When the tide is rising and the flow of the current is directed towards the shore, the tidal current is called the flood current, and when the tide is receding and the current is directed back out to sea, it is called the ebb current. Because the relative positions of the moon, sun and earth change at a known rate, tidal currents are predictable.

Tidal Stream

Tidal Stream is the name given to the horizontal flow of water through the oceans caused by the continuous ebb and flood of the tide, which as we know is the vertical up-down movement of the oceans water. Unlike water currents which are a continuous, unidirectional and form a steady horizontal movement of water flowing down a river or stream etc. a *tidal stream* or tidal current, changes its speed, direction and horizontal movement regularly according to the forces of the tide controlling it.

Tidal stream generation is a non-barrage tidal scheme, unlike tidal fence energy which uses a physical barrier to extract the energy. Tidal stream systems extract the kinetic energy (energy in motion) from moving water generated by the tides without altering the environment thereby making it a hydrokinetic energy system.

At or near the coast, the ebb and flood of the tides causes the oceans waters to pile up resulting in a high tide along the beach, with some of this water being forced into tidal inlets, basins and estuaries while the majority is forced sideways along the shore. This movement of the tidal range amplified by geographical features along the coastline, focuses these tidal currents into a single predictable and concentrated form of renewable energy which we can exploit using a tidal stream generator. A tidal stream is usually stronger nearer to the coast where the sea water is naturally shallower causing the water to speed, than it is farther out in deeper depths.

Tidal Stream Generator.

Tidal Stream Generation is very similar in many ways to the principles of wind power generation. Horizontal turbine generators called "tidal turbines" or "marine current turbines" are placed on the ocean floor, the stream currents flow across the turbine blades powering a generator much like how wind turns the blades of wind power turbines. In fact, in some tidal stream generation areas the sea bed looks just like underwater wind farm with arrays of tidal stream generators covering large areas.

The generated tidal electricity is then transmitted to the shore via long underwater electrical cables called *submarine cables*. These offshore tidal turbines can be either partially or fully submerged beneath the surface of the water, with partially submerged turbines being easier and less costly for maintenance.

While tidal stream installations reduce some of the environmental effects of large man-made tidal barrages, major ocean currents like the Gulf Stream, travel at speeds significantly slower than the wind. However, as water is 784 times more dense than air (which is why we can see water and not air), a single tidal generator sitting on the sea bed can provide a significant amount of ocean current energy at low tidal stream velocities which is far superior to wind, using similar or identical turbine technology.

Since energy output varies with the density of the medium, (Kg/cm^3) and the cube of the velocity, (m^3/s), we can see that a 10 mph (about 8.6 knots in nautical terms) ocean tidal current would have an energy output equal or greater than a 90 mph wind speed for the same size of turbine system. Therefore, even small increases in velocity can lead to substantial changes in the amount of available power and therefore, smaller faster rotating tidal turbine generators can be used in a ocean based tidal stream system.

As the kinetic energy content of a tidal stream flows per unit time, which is the same as the hydro power (P), the available energy can be calculated in terms of velocity (V), swept cross-sectional area (A) perpendicular to the stream flow direction, and the density of the water (ρ), which for sea water is approximately 1025 kg/m^3. Providing the velocity is uniform across the cross-sectional area, at any instant in the tidal cycle the amount of energy available will be: $P = \frac{1}{2} \rho A.V^3$.

This cubic relationship between velocity and power is the same as that for the power curves relating to wind turbines, but there are practical limits to the amount of power that can be extracted from tidal streams. Some of these limits relate to the design of the tidal stream turbines and the characteristics of the underwater resource.

Tidal stream generators make use of the kinetic energy of moving water to turn a turbine - similar to the way a wind turbine uses wind to create electricity. However, the power available for tidal power generation in a given area can be greater than a wind turbine due to the higher density of water. These types of tidal generators tend to be the cheapest (though still quite expensive) and most environmentally friendly of any type of tidal power generation. There are several different specific types of tidal stream generators.

These types of generators have a very low visual impact and are mainly, if not totally submerged. Furthermore, they are less intrusive to marine life because they produce less noise pollution.

Tides are more predictable, than wind which varies a great deal. The density of water is also much higher than air, which means that tidal turbines can be much smaller than wind turbines of comparable output. This is further enhanced by the use of the Venturi effect, which is a way of getting water to move faster through these turbines.

As with all types of tidal power technology the initial investment costs are extremely high and it takes quite a while to make back that investment. Unfortunately, sea water is quite corrosive which leads to high maintenance costs.

Since the tides flow-in and recede once each per day tidal stream generators provide intermittent power generation.

Varying types of generators

Different designs of tidal power generation may prove to offer different advantages, extensive studies are still needed on this relatively new technology. Studies are needed for maintenance costs vs. income and and environmental impacts over an extended period of time. Power output can vary considerably between these classes of turbines. However, since the tidal power industry is so new predictions for power output vary considerably. Currently, research data on stream generators is scarce due to the emerging nature of the industry.

Axial (Horizontal) Turbines

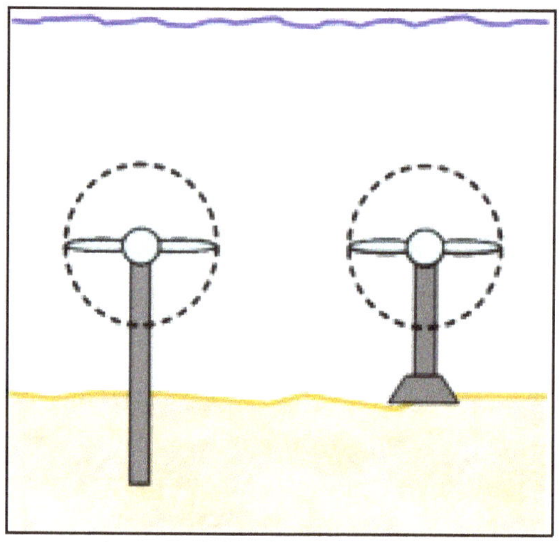

Figure: Axial turbines underwater.

Axial turbines have a rotor that is parallel to the incoming water stream. They use rotors similar to those seen on wind turbines but have modifications due to the differing fluid properties of water from air.

These devices use the lift of water to generate power. This requires specially shaped airfoil surfaces designed to create a pressure difference. This leads to a net force in the direction perpendicular to the water flow and thus turns the device. Rotors of this type must be carefully oriented (the orientation is referred to as the rotor pitch), to maintain their ability to harness the power of the tide as it changes.

Cross-flow Turbines

Cross flow turbines have a rotor that is perpendicular to the water flow but parallel to the water surface, see figure above Unlike axial turbines, these devices typically use the drag of water to generate power. In drag-based turbines, the force of the water pushes against a surface, like wind on an open sail. This works because the drag of the open face of the turbine blade is greater than the drag on its closed face. Drag based devices are inherently less efficient than their lift based counterparts because they operate with respect to the relative flow speed of the water. This means that as a turbine rotates faster the relative flow speed of the water will decrease and thus transfer less of its energy.

Figure: A cross flow water turbine.

Vertical axis Turbines

With vertical axis turbines the rotational axis of the rotor is vertical to the water surface and also perpendicular to the incoming water flow. Both lift and drag type blades can be used in these turbines.

Oscillating Generators

Oscillating devices do not have a rotational component, instead these generators make use of sections which are pushed sideways by the flowing current to create a hydraulic pump. This pump transfers its energy to a motor, which then turns a generator and creates electricity.

Tidal Stream Generator Designs

Unlike off-shore wind power which can suffer from storm or heavy sea damage, tidal or marine current turbines operate just below the sea surface or are permanently fixed to the sea bed. Most submerged tidal turbines essentially operate in the same way as a wind turbine and are fastened to the ocean floor, with water pushing the turbine instead of the wind. These turbines have an axis of rotation horizontal to the ground and operates like a traditional windmill consisting of a rotor, a gearbox, and an electrical generator. These three parts are mounted onto a steel support structure with the three main types of support being a gravity structure, a sunken piled structure or a tripod structure as shown,

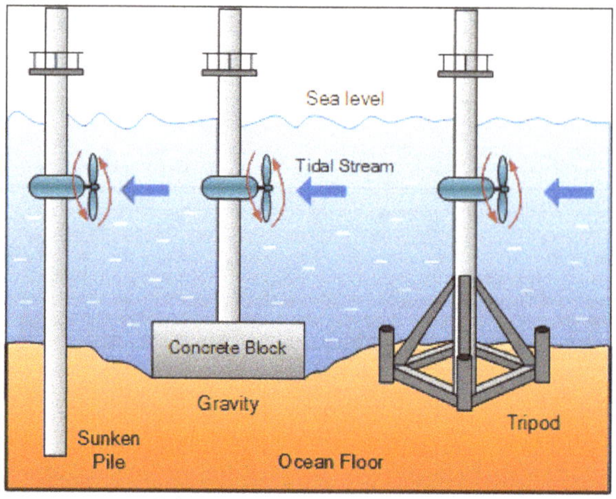

Tidal Stream Generator Supports.

For a sunken pile support, a single steel pile is driven deep into the sea bed with the tidal stream generator assembly attached to it. This tubular support is less stiff than other types and can flex under the downstream drag forces of the tidal waters when used in shallow waters. A gravity support generally uses a large heavy concrete block or blocks which sit on the sea bed. Due to the heavy weight of the concrete block, the structure is stiffer and therefore more resistant to flexing. A tripod or truss support uses a tubular frame with a much larger footprint positioned on the ocean floor to support the generator assembly. This type of system is used in oil and gas exploration so is a known technology.

Other tidal stream generator designs fixed to the ocean floor include: Reciprocating Tidal Stream Devices that uses a large hydrofoil similar to a whales flipper, which moves up and down parallel to the direction of the tidal stream instead of rotating blades, and Venturi Effect Tidal Stream Devices, were the tidal turbines are located inside a cylindrical duct, much like a fan housing. The tidal flow is funnelled through this duct, which concentrates the flow producing a pressure difference causing a secondary water flow through the reaction turbine thereby improving efficiency.

Also, there are several practical advantages in placing the tidal turbine inside a fan type duct, such as less dangers from the rotating blades to both aquatic marine life and divers as a safety grill or cover could be placed on the upstream opening which would also have the secondary advantage of preventing floating debris from being drawn or sucked into the turbine causing damage. The duct itself can provide shading and/or shelter for the reaction turbine from direct sunlight, preventing seaweed, algae growth or crustaceans forming on the blades and mechanism as they do on the underside of boats.

We know that tidal streams are formed by the fast flowing horizontal currents of water caused by the ebb and flow of the tides with the profile of the sea bed causing the water currents to speed up, or slow down near the shoreline. Then tidal stream turbines can generate power on both the ebb and flow of the tide. One of the disadvantages of Tidal

Stream Generation is that, as the turbines are submerged under the surface of the water they can create hazards to large sea mammals, navigation and shipping.

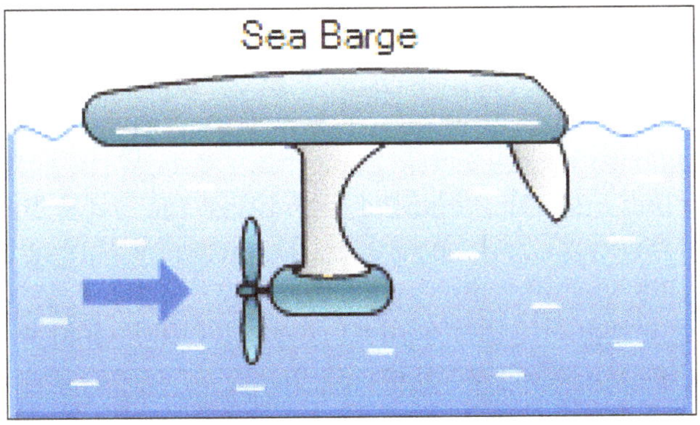

Given the technical difficulties resulting from underwater corrosion, increased maintenance issues, weed growth on the blades, which could reduce their efficiency and stability concerns, other forms of alternative tidal stream generator designs are now being used. These include the tidal turbine being connected to a floating barge or ship on the waters surface, essentially operating as an upside down horizontal turbine instead of fastening the turbines directly to the ocean floor.

There are numerous advantages to this type of tidal stream generator design, including easy maintenance and accessibility of the turbines, by simply removing or replacing them out of the water, and no costly steel supports or alterations to the ocean floor. Also, as the tidal turbines are located under a barge, pontoon or fixed directly to the hull of a ship, they can have their electrical connections and equipment mounted safely above and out of the water. Plus the supporting flotation device can be easily moved to stronger tidal stream areas if required, but they are limited by distance due to their umbilical electrical cable connected to the shoreline.

Tidal Barrage

The Tidal Barrage or Tidal Power Plant as it is also known is a form of "marine renewable energy" generation system that uses long walls, dams, sluice gates or tidal locks to capture and store the potential energy of the ocean. A *Tidal Barrage* is a type of tidal power generation scheme that involves the construction of a fairly low walled dam, known as a "tidal barrage" and hence its name, spanning across the entrance of a tidal inlet, basin or estuary creating a single enclosed tidal reservoir, similar in many respects to a hydroelectric impoundment reservoir.

The bottom of this barrage dam is located on the sea floor with the top of the tidal barrage being just above the highest level that the water can get too at the highest annual tide. The barrage has a number of underwater tunnels cut into its width allowing the sea water to flow through them in a controlled way by using "sluice gates" on their entrance and exit points. Fixed within these tunnels are huge tidal turbine generators that spin as the sea water rushes past them either to fill or empty the tidal reservoir thereby generating electricity.

The water which flows into and out of these underwater tunnels carries enormous amounts of kinetic energy and the job of the tidal barrage is to extract as much of this energy as possible which it uses to produce electricity. Tidal barrage generation using the tides is very similar to hydroelectric generation, except the water flows in two directions rather than just one. On incoming high tides, the water flows in one direction and fills up the tidal reservoir with sea water. On outgoing ebbing tides, the sea water flows in the opposite direct emptying it. As a tide is the vertical movement of water, the tidal barrage generator exploits this natural rise and fall of tidal waters caused by the gravitational pull of the sun and the moon.

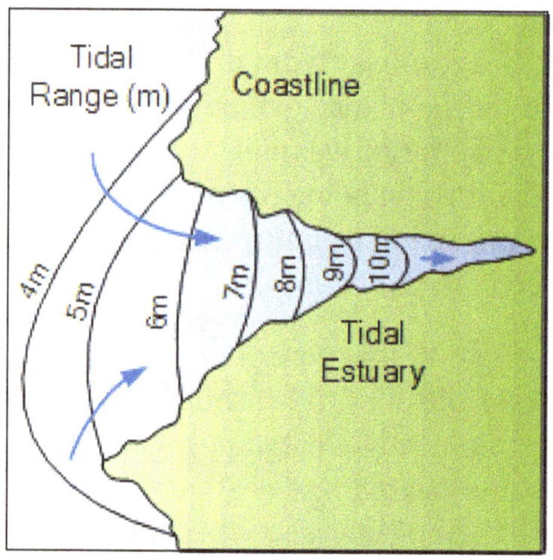

Effects of Tidal Flow in an Estuary.

The gravitational effects of the sun or the moon on the worlds oceans causes huge amounts of sea water to be directed towards the nearest coastline. The result of this movement of water is a rise in the sea level. In the open ocean, this rise is very small as there is a large surface area with deeper depths for it to flow into.

However, as the oceans water moves nearer towards the coastline, the sea level rises steeply especially around inlets and estuaries because of the upward sloping gradient of the sea bed. The effect of this sloping gradient is to funnelling the water into the estuaries, lagoons, river inlets and other such tidal "bottlenecks" along the coastline.

The result of funnelling all of this water is that the height of the sea level once inside these inlets can increase vertically many metres every day as it is being pushed forward by the incoming sea water behind it as shown in the image. This increase in the sea level can create a tidal range of over ten metres in height in some estuaries and locations which can be exploited to generate electricity.

The tidal range is the vertical difference between the high tide sea level and the low tide sea level. The tidal energy extracted from these tides is potential energy as the tide moves in a vertical up-down direction between a low and a high tide and back to a low creating a height or head differential. A tidal barrage generation scheme exploits this head differential to generate electricity by creating a difference in the water levels either side of a dam and then passing this water difference through the turbines. The three main tidal energy barrage schemes that use this water differential to their advantage are:

- Flood Generation: in which the tidal power is generated as the water enters the tidal reservoir on the incoming tide.
- Ebb Generation: in which the tidal power is generated as the water leaves the tidal reservoir on the ebb tide.
- Two-way Generation: in which the tidal power is generated as the water flows in both directions during a flood and ebb tide.

Tidal Barrage Flood Generation

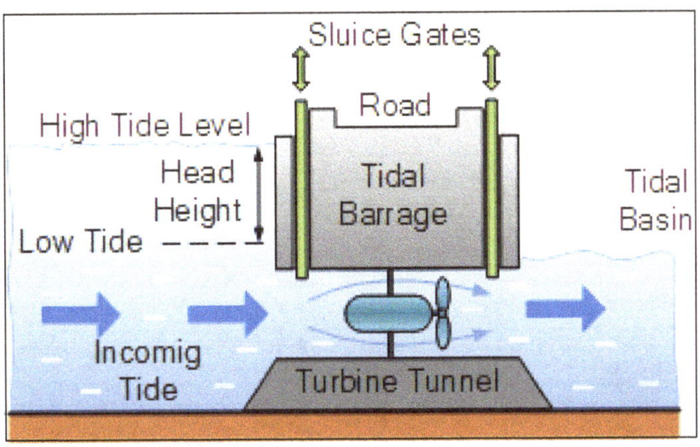

A Tidal Barrage Flood Generation uses the energy of an incoming rising tide as it moves towards the land. The tidal basin is emptied through sluice gates or lock gates located along a section of the barrage and at low tide the basin is affectively empty. As the tide turns and starts to comes in, the sluice gates are closed and the barrage holds back the rising sea level, creating a difference in height between the levels of water on either side of the barrage dam.

The sluice gates to the entrances to the dams tunnels can either be closed as the sea water rises to allow for a sufficient head of water to develop between the sea level and the basin level before being opened generating more kinetic energy as the water rushes through, turning the turbines as it passes. Or may remain fully open, filling up the basin more slowly and maintaining the same water level inside the basin as out in the sea.

The tidal reservoir is therefore filled up through the turbine tunnels which spin the turbines generating tidal electricity on the flood tide and is then emptied through the opened sluice or lock gates on the ebb tide. Then a flood tidal barrage scheme is a one-way tidal generation scheme on the incoming tide with tidal generation restricted to about 6 hours per tidal cycle as the basin fills up.

The movement of the water through the tunnels as the tidal basin fills up can be a slow process, so low speed turbines are used to generate the electrical power. This slow filling cycle allows for fish or other sea life to enter the enclosed basin without danger from the otherwise fast rotating turbine blades. Once the tidal basin is full of water at high tide, all the sluice gates are opened allowing all the trapped water behind the dam to return back to the ocean or sea as it ebbs away.

Flood generator tidal power generates electricity on a incoming or flood tide, but this form of tidal energy generation is generally much less efficient than generating electricity as the tidal basin empties, called "Ebb Generation". This is because the amount of kinetic energy contained in the lower half of the basin in which flood generation operates is much less the kinetic energy present in the upper half of the basin in which ebb generation operates due to the effects of gravity and the secondary filling of the basin from inland rivers and streams connected to it via the land.

Tidal Barrage Ebb Generation

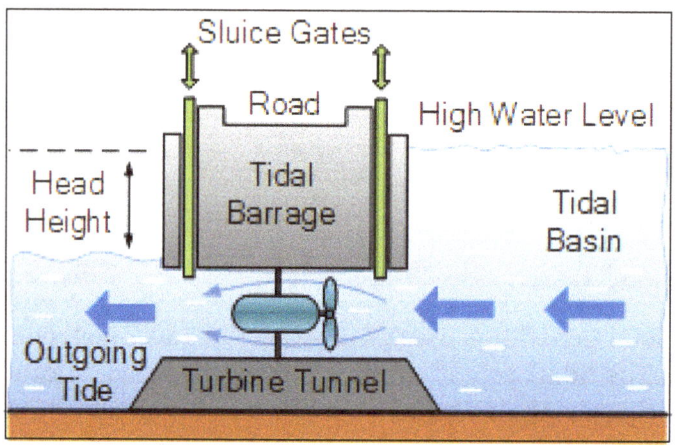

A Tidal Barrage Ebb Generation uses the energy of an outgoing or falling tide, referred to as the "ebb tide", as it returns back to the sea making it the opposite of the previous flood tidal barrage scheme. At low tide, all the sluice and lock gates along the barrage are fully opened allowing the tidal basin to fill up slowly at a rate determined by the incoming flood tide.

When the ocean or sea level feeding the basin reaches its highest point at high tide, all the sluices and lock gates are then closed entrapping the water inside the tidal basin (reservoir). This reservoir of water may continue to fill-up due to inland rivers and streams connected to it from the land.

As the level of the ocean outside the reservoir drops on the outgoing tide towards its low tide mark, a difference between the higher level of the entrapped water inside the tidal reservoir and the actual sea level outside now exists. This difference in vertical height between the high level mark and the low mark is known as the "head height".

At some time after the beginning of the ebb tide, the difference in the head height across the tidal barrage between the water inside the tidal reservoir and the falling tide level outside becomes sufficiently large enough to start the electrical generation process and the sluice gates connected to the turbine tunnels are opened allowing the water to flow.

When the closed sluice gates are opened, the trapped potential energy of the water inside flows back out to the sea under the enormous force of both gravity and the weight of the water in the reservoir basin behind it. This rapid exit of the water through the tunnels on the outgoing tide causes the turbines to spin at a fast speed generating electrical power.

The turbines continue to generate this renewable tidal electricity until the head height between the external sea level and the internal basin is too low to drive the turbines at which point the turbines are disconnected and the sluice gates closed again to prevent the tidal basin from over draining and effecting local wildlife. At some point the incoming flood tide level will again be at a sufficient level to open all the lock gates filling-up the basin and repeating the whole generation cycle over again as shown.

Power Generation during Ebb Tide

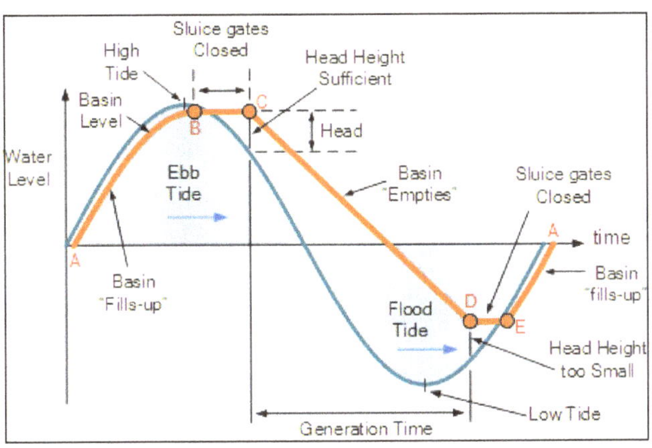

From the graph above we can see that the tidal basin fills up between points E and B via A on the incoming flood tide. Once high tide is reached, the sluice gates are then closed between points B – C to stop the tidal basin from emptying. Therefore no power is generated between points E – C while the basin is filling up.

When there is sufficient head height between either side of the barrage, the sluice gates are opened at point C releasing the trapped water back to the sea and power generation starts. The tidal basin continues to empty driving the tidal generators until the level of water in the basin reaches point D.

At point D the head height of the sea water across the barrage is no longer adequate to drive the turbines so the sluice gates are closed and generation stops until the level of the sea water reaches point E again and the whole process starts over. Then Ebb Tidal Barrage Generation also known as outgoing generation, takes its name because the electrical generation only occurs as the tide ebbs or flows out.

At point E the sea level becomes sufficient to re-fill the basin and the sluice gates are opened. Therefore, the tidal reservoir is filled up through the sluice or lock gates on the flood tide and is emptied through the turbine tunnels which spin the turbines generating tidal electricity on the ebb tide. Then an ebb tidal barrage scheme is a one-way tidal generation scheme which operates between points C – D on the graph above.

Two-way Tidal Barrage Generation Scheme

We have seen above that both Flood Tidal Barrage and Ebb Tidal Barrage installations are "one-way" tidal generation schemes, but in order to increase the power generation time and therefore improve efficiency, we can use special double effect turbines that generate power in both directions. A Two-way Tidal Barrage Scheme uses the energy over parts of both the rising tide and the falling tide to generate electricity.

Two-way electrical generation requires a more accurate control of the sluice gates keeping them closed until the differential head height is sufficient in either direction before being opened. As the tide ebbs and flows, sea water flows in or out of the tidal reservoir through the same gate system. This flow of tidal water back and forth causes the turbine generators located within the tunnel to rotate in both directions producing electricity.

However, this two-way generation is in general less efficient than one-way flood or ebb generation as the required head height is much smaller which reduces the period over which normal one-way generation might have otherwise occurred. Also, bi-directional tidal turbine generators designed to operate in both directions are generally more expensive and less efficient than dedicated unidirectional tidal generators.

One way of improving the operating time and efficiency of a two-way tidal barrage scheme is to use individual one-way unidirectional tidal turbines inverted along the barrage. By controlling their individual sluice gates, one set of turbines can be made to operate on the flood tide between points E - B and the other set operate on the ebb tide between points C - D. While this method increases the total number of tidal turbines located along the tidal barrage, it has the advantage that the generation period is greatly extended.

Pumping

Turbines are able to be powered in reverse by excess energy in the grid to increase the water level in the basin at high tide (for ebb generation). Much of this energy is returned during generation, because power output is strongly related to the head. If water is raised 2 ft (61 cm) by pumping on a high tide of 10 ft (3 m), this will have been raised by 12 ft (3.7 m) at low tide.

Environmental Impact

The placement of a barrage into an estuary has a considerable effect on the water inside the basin and on the ecosystem. Many governments have been reluctant in recent times to grant approval for tidal barrages. Through research conducted on tidal plants, it has been found that tidal barrages constructed at the mouths of estuaries pose similar environmental threats as large dams. The construction of large tidal plants alters the flow of saltwater in and out of estuaries, which changes the hydrology and salinity and possibly negatively affects the marine mammals that use the estuaries as their habitat. The La Rance plant, off the Brittany coast of northern France, was the first and largest tidal barrage plant in the world. It is also the only site where a full-scale evaluation of the ecological impact of a tidal power system, operating for 20 years, has been made.

French researchers found that the isolation of the estuary during the construction phases of the tidal barrage was detrimental to flora and fauna, however; after ten years, there has been a "variable degree of biological adjustment to the new environmental conditions".

Some species lost their habitat due to La Rance's construction, but other species colonized the abandoned space, which caused a shift in diversity. Also as a result of the construction, sandbanks disappeared, the beach of St. Servan was badly damaged and high-speed currents have developed near sluices, which are water channels controlled by gates.

Turbidity

Turbidity (the amount of matter in suspension in the water) decreases as a result of smaller volume of water being exchanged between the basin and the sea. This lets light from the Sun penetrate the water further, improving conditions for the phytoplankton. The changes propagate up the food chain, causing a general change in the ecosystem.

Tidal Fences and Turbines

Tidal fences and turbines, if constructed properly, pose less environmental threats than tidal barrages. Tidal fences and turbines, like tidal stream generators, rely entirely on the kinetic motion of the tidal currents and do not use dams or barrages to block channels or estuarine mouths. Unlike barrages, tidal fences do not interrupt fish migration or alter hydrology, thus these options offer energy generating capacity without dire environmental impacts. Tidal fences and turbines can have varying environmental impacts depending on whether or not fences and turbines are constructed with regard to the environment. The main environmental impact of turbines is their impact on fish.

If the turbines are moving slowly enough, such as low velocities of 25-50 rpm, fish kill is minimalized and silt and other nutrients are able to flow through the structures For example, a 20 kW tidal turbine prototype built in the St. Lawrence Seaway in 1983 reported no fish kills Tidal fences block off channels, which makes it difficult for fish and wildlife to migrate through those channels. In order to reduce fish kill, fences could be engineered so that the spaces between the caisson wall and the rotor foil are large enough to allow fish to pass through Larger marine mammals such as seals or dolphins can be protected from the turbines by fences or a sonar sensor auto-braking system that automatically shuts the turbines down when marine mammals are detected.

Salinity

As a result of less water exchange with the sea, the average salinity inside the basin decreases, also affecting the ecosystem. "Tidal Lagoons" do not suffer from this problem.

Sediment movements

Estuaries often have high volume of sediments moving through them, from the rivers to the sea. The introduction of a barrage into an estuary may result in sediment accumulation within the barrage, affecting the ecosystem and also the operation of the barrage.

Fish

Fish may move through sluices safely, but when these are closed, fish will seek out turbines and attempt to swim through them. Also, some fish will be unable to escape the water speed near a turbine and will be sucked through. Even with the most fish-friendly turbine design, fish mortality per pass is approximately 15% (from pressure drop, contact with blades, cavitation, etc.). Alternative passage technologies (fish ladders, fish lifts, fish escalators etc.) have so far failed to solve this problem for tidal barrages, either offering extremely expensive solutions, or ones which are used by a small fraction of fish only. Research in sonic guidance of fish is ongoing. The Open-Centre turbine reduces this problem allowing fish to pass through the open centre of the turbine.

Recently a run of the river type turbine has been developed in France. This is a very large slow rotating Kaplan-type turbine mounted on an angle. Testing for fish mortality has indicated fish mortality figures to be less than 5%. This concept also seems very suitable for adaption to marine current/tidal turbines.

Energy calculations

The energy available from a barrage is dependent on the volume of water. The potential energy contained in a volume of water is:

$$E=\tfrac{1}{2}A\rho g h^2$$

Where:

- h is the vertical tidal range,
- A is the horizontal area of the barrage basin,
- ρ is the density of water = 1025 kg per cubic meter (seawater varies between 1021 and 1030 kg per cubic meter),
- g is the acceleration due to the Earth's gravity = 9.81 meters per second squared.

The factor half is due to the fact, that as the basin flows empty through the turbines, the hydraulic head over the dam reduces. The maximum head is only available at the moment of low water, assuming the high water level is still present in the basin.

Example Calculation of Tidal Power Generation

Assumptions:

- The tidal range of tide at a particular place is 32 feet = 10 m (approx).
- The surface of the tidal energy harnessing plant is 9 km^2 (3 km × 3 km)= 3000 m × 3000 m = 9×10^6 m^2
- Density of sea water = 1025.18 kg/m^3

Mass of the sea water = volume of sea water × density of sea water.

= (area × tidal range) of water × mass density

= $(9 \times 10^6 \text{ m}^2 \times 10 \text{ m}) \times 1025.18 \text{ kg/m}^3$

= 92×10^9 kg (approx)

Potential energy content of the water in the basin at high tide = ½ × area × density × gravitational acceleration × tidal range squared.

= $\frac{1}{2} \times 9 \times 10^6 \text{ m}^2 \times 1025 \text{ kg/m}^3 \times 9.81 \text{ m/s}^2 \times (10 \text{ m})^2$

=4.5×10^{12} J (approx)

Now we have 2 high tides and 2 low tides every day. At low tide the potential energy is zero. Therefore, the total energy potential per day = Energy for a single high tide × 2.

= 4.5×10^{12} J × 2

= 9×10^{12} J

Therefore, the mean power generation potential = Energy generation potential / time in 1 day.

$= 9 \times 10^{12}$ J / 86400 s

= 104 MW

Assuming the power conversion efficiency to be 30%: The daily-average power generated = 104 MW * 30%.

= 31 MW (approx)

Because the available power varies with the square of the tidal range, a barrage is best placed in a location with very high-amplitude tides. Suitable locations are found in Russia, USA, Canada, Australia, Korea, the UK. Amplitudes of up to 17 m (56 ft) occur for example in the Bay of Fundy, where tidal resonance amplifies the tidal range.

Wave Power

Wave Energy also known as Ocean Wave energy, is a type of ocean based renewable energy source that uses the power of the waves to generate electricity.

Unlike tidal energy which uses the ebb and flow of the tides, *wave energy* uses the vertical movement of the surface water that produce tidal waves. Wave power converts the periodic up-and-down movement of the oceans waves into electricity by placing equipment on the surface of the oceans that captures the energy produced by the wave movement and converts this mechanical energy into electrical power.

Wave energy is actually a concentrated form of solar power generated by the action of the wind blowing across the surface of the oceans water which can then be used as a renewable source of energy. As the sun rays strike the Earth's atmosphere, they warm it up. Differences in the temperature of the air masses around the globe causes the air to move from the hotter regions to the cooler regions, resulting in winds.

As the wind passes over the surface of the oceans, a portion of the winds kinetic energy is transferred to the water below, generating waves. In fact, the ocean could be viewed as a vast storage collector of energy transferred by the sun to the oceans, with the waves carrying the transferred kinetic energy across the surface of the oceans. Then we can say that waves are actually a form of energy and it is this energy and not water that moves along the ocean's surface.

These waves can travel (or "propagate") long distances across the open oceans with very little loss in energy, but as they approach the shoreline and the depth of the water becomes shallower, their speed slows down but they increase in size. Finally, the wave crashes onto the shoreline, releasing an enormous amount of kinetic energy which can be used for electricity production. A breaking waves energy potential varies from place to place depending upon its geographic location and time of year, but the two main factors which affect the size of the wave energy are the winds strength and the uninterrupted distance over the sea that the wind can blow.

Then we can say that "Wave Energy" is an indirect form of wind energy that causes movement of the water on the surface of the oceans and by capturing this energy the motion of the waves is converted to mechanical energy and used to drive an electricity generator. In many respects, the technology used for capturing this wave energy is similar to tidal energy or hydroelectric power.

The kinetic energy of the wave turns a turbine attached to a generator, which produces electricity. However, the open oceans can be a stormy and violent environment, resulting in the wave energy machines being destroyed by the very energy they were designed to capture.

In its simplest terms, an ocean wave is the up-and-down vertical movement of the sea water which varies sinusoidally with time. This sinusoidal wave has high points called crests and low points called troughs. The difference in height of a wave between the crest and the trough is called the peak-to-peak amplitude, then the waves amplitude or height is the centre of these two points and corresponds to the actual sea level when there is no movement of the water, in other words, a calm sea.

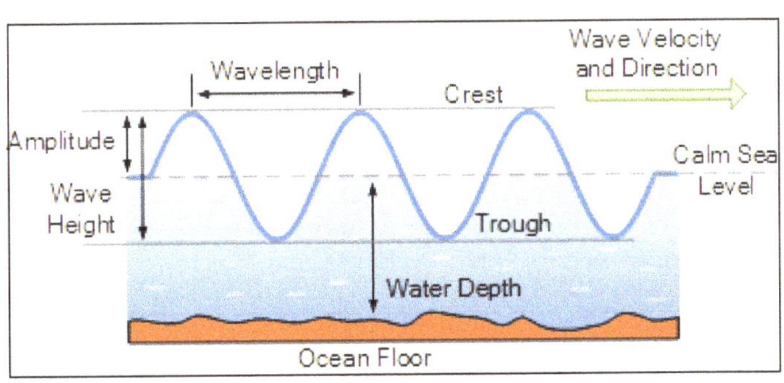

The amplitude of an ocean wave depends on the weather conditions at that time, as the amplitude of a smooth wave, or swell, will be small in calm weather but much larger in stormy weather with strong gales as the sea water moves up and down.

As well as the amplitude of the wave, another important characteristic is the distance between each successive crest, or trough, known as the wave period, (T). This wave period is the time in seconds between each crest of the wave. Then for a gentle swell this time period may be very long, but for a stormy sea this time period may be very short as each wave crashes onto the one in front.

The reciprocal of this time (1/T) gives us the fundamental frequency of the ocean wave relative to some static point. Smaller periodic waves generated or superimposed onto this fundamental wave such as reflected waves are called harmonic waves. Then the frequency and amplitude characteristics of a wind-generated wave depend on the distance the wind blows over the open water (called the fetch), the length of time the wind blows, the speed of the wind and the water depth.

Waves transport energy from where they were created by storms far out in the ocean to a shoreline. But a typical ocean wave does not resemble a perfect sinusoid, they are more irregular and complex than a simple sinusoidal wave. Only the steady up-and-down movement of a heavy swell resembles a sinusoidal wave much more than the chaotic nature of locally generated wind waves, as real sea waves contain a mixture of waves with different frequencies, wave heights and directions.

Wave Power Devices

Ocean wave energy has many advantages over ocean wind energy in that it is more predictable, less variable and offers higher available energy densities. Depending on the distance between the energy conversion device and the shoreline, wave energy systems can be classified as being either *Shoreline devices, Nearshore devices* or *Offshore devices.* So what is the difference between these three types of energy extraction devices.

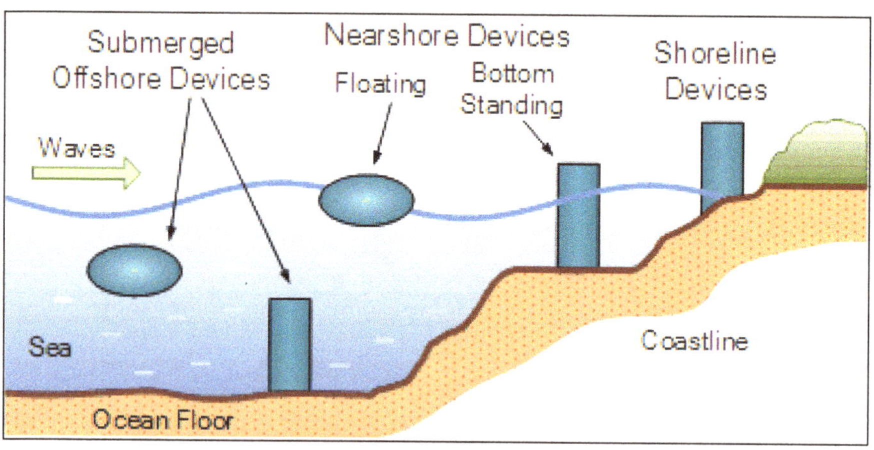

Shoreline devices are wave energy devices which are fixed to or embedded in the shoreline, that is they are both in and out of the water. Nearshore devices are characterised by being used to extract the wave power directly from the breaker zone and the waters immediately beyond the breaker zone, (i.e. at 20m water depth).

Offshore devices or deep water devices are the farthest out to sea and extend beyond the breaker lines utilising the high-energy densities and higher power wave profiles available in the deep water waves and surges.

One of the advantages of offshore devices is that there is no need for significant coastal earthworks, as there is with onshore devices.

As most of the energy within a wave is contained near the surface and falls off sharply with depth. There is a surprising range of designs available that maximise the energy available for capture. These wave energy devices are either fixed bottom standing designs used in shallow water and which pierce the water surface, or fully floating devices that are used to capture the kinetic energy content of a waves movement and convert each movement into electricity using a generator.

There are currently four basic "capture" methods:

- Point Absorbers: These are small vertical devices either fixed directly to the ocean floor or tethered via a chain that absorb the waves energy from all directions. These devices generate electricity from the bobbing or pitching action of a floating device. Typical wave energy devices include, floating buoys, floating bags, ducks, and articulated rafts, etc. These devices convert the up-and-down pitching motion of the waves into rotary movements, or oscillatory movements in a variety of devices to generate electricity. One of the advantages of floating devices over fixed devices it that they can be deployed in deeper water, where the wave energy is greater.
- Wave Attenuators: These are also known as "linear absorbers", are long horizontal semi-submerged snake-like devices that are oriented parallel to the direction of the waves. A wave attenuator is composed of a series of cylindrical sections linked together by flexible hinged joints that allow these individual sections to rotate and yaw relative to each other. The wave-induced motion of the device is used to pressurise a hydraulic piston, called a ram, which forces high pressure oil through smoothing accumulators to turn a hydraulic turbine generator producing electricity. Then wave attenuators convert the oscillating movement of a wave into hydraulic pressure.
- Oscillating Water Column: It is a partly submerged chamber fixed directly at the shoreline which converts wave energy into air pressure. The structure could be a natural cave with a blow hole or a manmade chamber or duct with an wind turbine generator located at the top well above the waters surface. The structure is built perpendicular to the waves so that the ebbing and flowing motion of the waves force the trapped water inside the chamber to oscillate in the vertical direction.

As the waves enter and exit the chamber, the water column moves up and down and acts like a piston on the air above the surface of the water, pushing it back and forth. This air is compressed and decompressed by this movement and is channelled through a wind turbine generator to produce electricity. The speed of air in the duct can be enhanced by making the cross-sectional area of the duct much less than that of the column.

- Overtopping Devices: It is also known as "spill-over" devices, are either fixed or floating structures that use ramps and tapered sides positioned perpendicular to the waves. The sea waves are driven up the ramp and over the sides filling-up a small tidal reservoir which is located 2 to 3 metres above sea level. The potential energy of the water trapped inside the reservoir is then extracted by returning the water back to the sea through a low head Kaplan turbine generator to produce electricity.

Then overtopping devices convert the potential energy available in the head of water into mechanical energy. The disadvantage of onshore overtopping schemes is that they have a relatively low power output and are only suitable for sites where there is a deep water shoreline and a low tidal range of less than about a metre.

The idea of harnessing the tremendous power of the oceans waves is not new. Like other forms of hydro power, wave energy does not require the burning of fossil fuels, which can pollute the air, contributing to acid rain and global warming. The energy is entirely clean and endlessly renewable. Wave power has many advantages compared to other forms of renewable energy with its main advantage being that it is predictable.

However, like many other forms of renewable energy, ocean wave energy also has its disadvantages such as its inflexible generation times dependant upon the tides, the visual impact of wave devices on the seas surface, as well as the threat of collision to shipping and navigation.

Here are some of the main advantages and disadvantages of wave energy:

Wave Energy Advantages

- Wave energy is an abundant and renewable energy resource as the waves are generated by the wind.
- Pollution free as wave energy generates little or no pollution to the environment compared to other green energies.
- Reduces dependency on fossil fuels as wave energy consumes no fossil fuels during operation.
- Wave energy is relatively consistent and predictable as waves can be accurately forecast several days in advance.
- Wave energy devices are modular and easily sited with additional wave energy devices added as needed.

- Dissipates the waves energy protecting the shoreline from coastal erosion.
- Presents no barriers or difficulty to migrating fish and aquatic animals.

Wave Energy Disadvantages

- Visual impact of wave energy conversion devices on the shoreline and offshore floating buoys or platforms.
- Wave energy conversion devices are location dependent requiring suitable sites were the waves are consistently strong.
- Intermittent power generation as the waves come in intervals does not generate power during calm periods.
- Offshore wave energy devices can be a threat to navigation that cannot see or detect them by radar.
- High power distribution costs to send the generated power from offshore devices to the land using long underwater cables.
- They must be able to withstand forces of nature resulting in high capital, construction and maintenance costs.

Run-of-River Hydroelectricity

Figure: Run-of-the-river systems like the one shown above tend to have larger hydroelectric flow rates than hydro dams that use reservoirs.

Run-of-the-river hydroelectric systems are hydroelectric systems that harvest the energy from flowing water to generate electricity in the absence of a large dam and reservoir—which is how they differ from conventional impoundment hydroelectric facilities. A small dam may be used to ensure enough water goes in the penstock, and possibly some storage (for same day use) The primary difference between this type of hydroelectric generation compared to others is that run-of-the-river primarily uses the natural flow rate of water to generate power—instead of the power of water falling a large distance. However, water may still experience some vertical drop in a run-of-the-river system from the natural landscape or small dam Another main difference between traditional hydropower is that run-of-the-river hydro is used in areas where there is little to no water storage, such as in a river.

There are several classifications of run-of-the-river systems, based primarily on their capacity. The types are outlined in the table below:

Classification	Capacity
Micro	< 100 kW
Mini	100 kW - 1 MW
Small	1 - 50 MW

It is important to note that some larger scale run-of-the-river plants exist, with outputs of hundreds or thousands of MW.

System Configurations

Run-of-river can mean several different things and have various configurations, as well as use different turbine technologies. Systems can involve the far upstream diversion of the river, diversion at an existing weir or dam, or simply in-stream current flow technology, and can involve penstock pipes, open channels, barrages and other diversion methods.

Penstock Based Systems or Streaming Systems

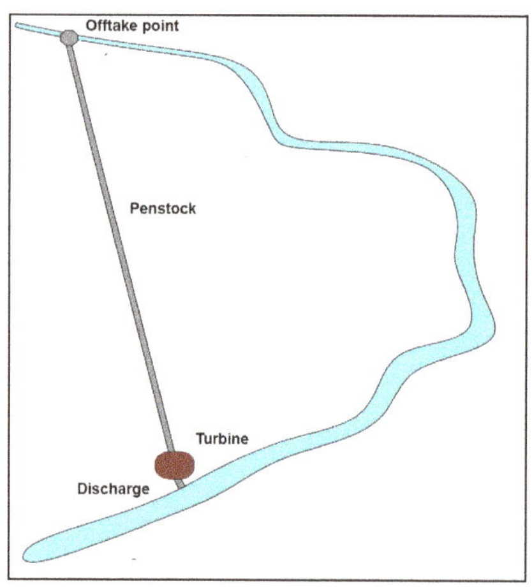

Figure: Penstock or streaming ROR system.

This diversion is known as the "depleted reach" and is measured as the distance between where the water leaves a river to enter a hydro system and where the water goes back into the river. Taking a given volume of water out of a river will have an impact upon the natural ecology of the river bed. A large depleted reach may therefore compromise the ecology of the river and its environment. The smaller the depleted distance, the less impact there will be upon the fish and wildlife ecology.

Diversion Systems

In these systems a short open canal or closed pipe is used to divert water from upstream of a low head artificial dam or natural drop, such as a waterfall or weir, into the turbine intake (Figure a). Typical examples are Inga 3 (Figure b), where a canal was constructed to divert the water flow upstream of the generator plant.

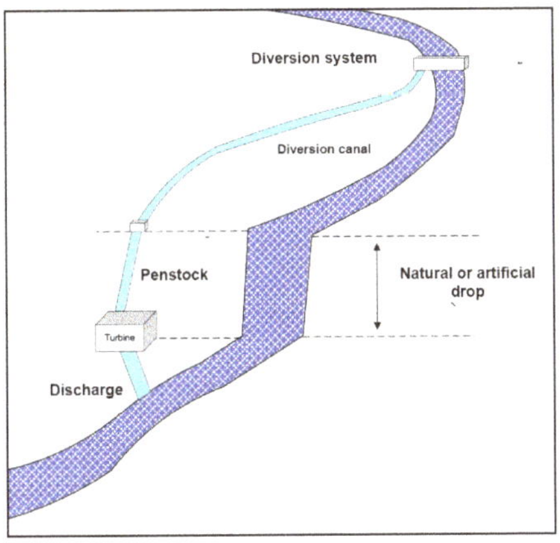

Figure a: River diversion system.

Figure b: Inga 3 hydropower system.

The drop across the Inga rapids or waterfall is approximately 96 m, although it is clear from figure b that the head created by the "dam" at Inga 3 is less than this. There is generally no limit on the amount of water which may be diverted. In many cases the turbine structure is incorporated into what amounts to a high weir or dam wall, the sole purpose of which is to feed water into the turbines.

Existing Dams and Weirs

ROR systems installed on existing dams use a diversion penstock around the wall of an existing dam with a low head, and significant inflow and outflow. Costs limit this approach to medium sized projects, i.e. the project must generate enough electricity to recover costs.

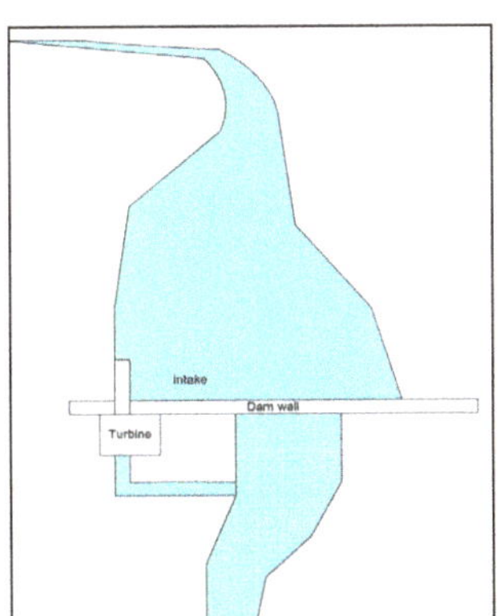

Figure: ROR system using existing dam.

Water is returned to the river. The capacity of the system will depend on the existing flow of water through the dam required to maintain the river flow. Diversion of water for the ROR scheme may replace the existing water release from the dam.

Barrage Systems

In this case a barrage is constructed across the river forming a low head. The turbines, in-line reaction impeller types such as Kaplan turbines, are mounted either in horizontal or vertical configuration in the barrage and are driven by the flow of the river water through the barrage. This type of system has been used extensively on large, wide rivers with a constant flow rate.

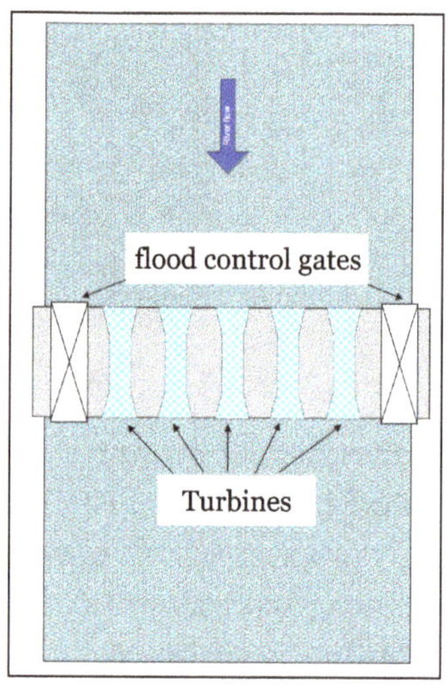

Figure: Barrage system.

Instream Systems (River Current Systems)

Limited at the moment to smaller systems, these use reaction turbines or small bulb turbines installed in the body of the river and extract energy from the river current, being a truly ROR system. There are several different types of turbine being used, from fully immersed turbines to vertical shaft turbines where the generator is not immersed. There is considerable interest in the development of these systems as they involve minimal construction of additional infrastructure and do not interfere with the flow of water in any way. Instream systems can make use of existing structures such as bridges.

ROR System Turbines

Turbines used with ROR systems may be impulse type or reaction type, and depend on the size of the system installed. Kaplan and Francis turbines are commonly used with kinetic systems, while Pelton wheel types are usually found in systems relying on a substantial head of water, such as the streaming system.

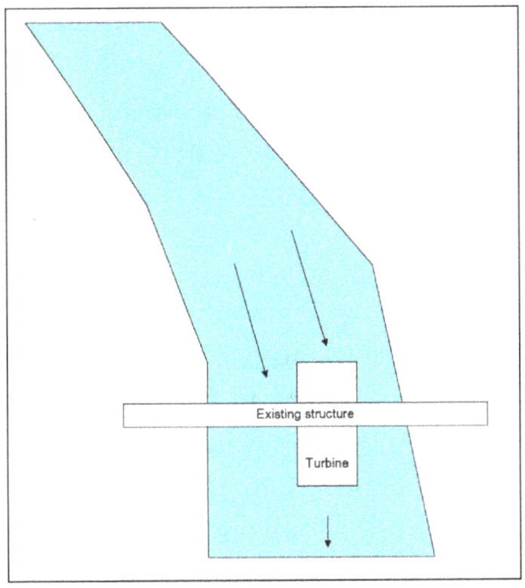

Figure: Instream systems.

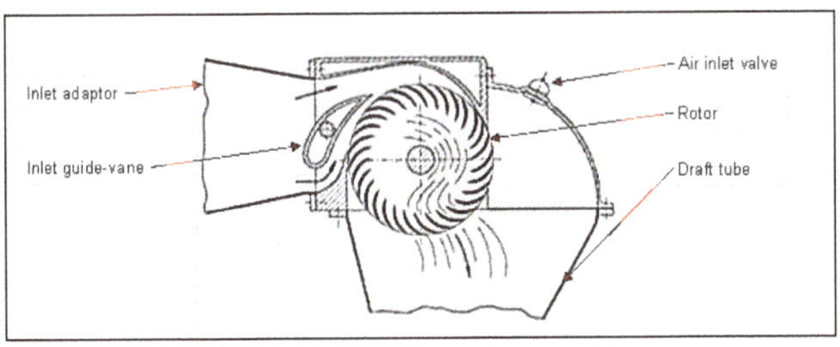

Figure: Vertical crossflow turbine.

Recent Developments in Turbines using ROR Techniques

Virtually a modern version of the waterwheel, the VCT is well adapted to ROR operation. Available in sizes ranging from tens of kW to several MW, this unit is essentially a micro-turbine, but could be applied easily to many of the smaller dams and water structures in the country. These turbines do not suffer damage from small debris in the water and are fairly rugged. The crossflow principle can be applied in either a horizontal inflow or a vertical inflow method.

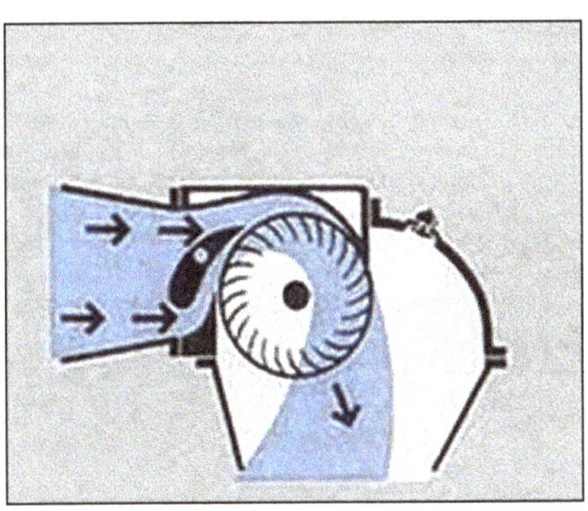

Figure a: Crossflow inflow horizontal.

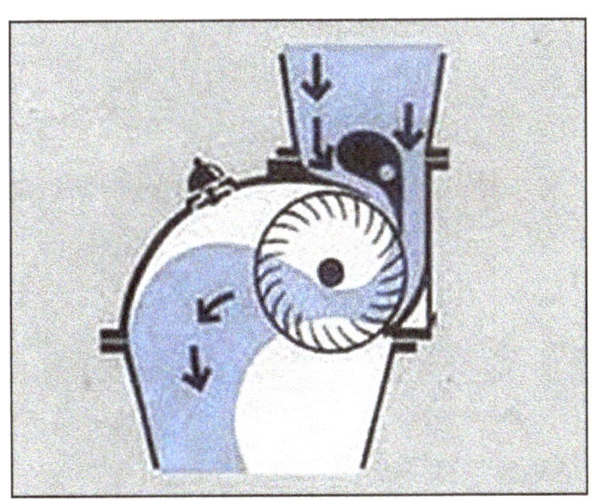

Figure b: Crossflow inflow vertical.

The main feature of the crossflow turbine is that water passes through the turbine blades twice. The first time on entry and the second time on exit after travelling through the centre of the blade wheel. Dual vertical crossflow systems, consisting of two crossflow turbines mounted vertically in a frame which spans the width of the feed canal have been used in some smaller installations.

Hydrofoil Kinetic Energy Turbine

A truly ROR turbine, kinetic energy turbines, also called free-flow turbines, generate electricity from the kinetic energy present in flowing water rather than the potential energy from the head. The systems may operate in rivers or man-made channels. Both vertical and horizontal axis versions are being developed. The vertical axis turbine can be placed in the river directly, and does not need a separate structure to control water flow. The vertical axis turbine has the advantage that the generator can be mounted above the turbine and therefore out of the water.

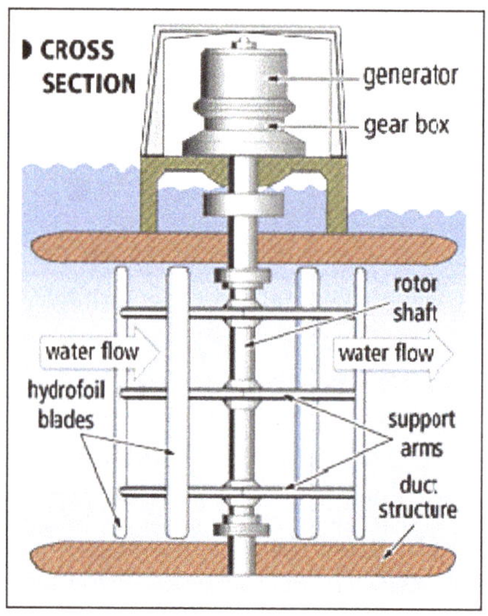

Figure a: Vertical axis turbine (Blue Energy Systems).

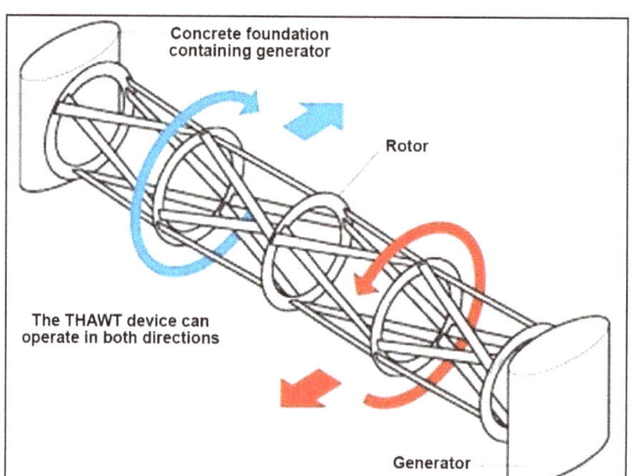

Figure b: The transverse horizontal axis water turbine.

Horizontal Axis "Bulb" Turbine

This is a more modern development using a Kaplan turbine for energy capture, the generator and control circuits are contained in a bulb and the whole device installed in a cone or cylinder to control water flow.

Archimedean Screw Turbines

These systems work on the principle of the Archimedean screw pump in reverse. . Sizes from 1 kW to 150 kW capacity are in service. The systems are fish friendly and are not susceptible to blocking by small debris. This is a true run-of-river device as the in feed takes place at the water surface, and the speed of rotation is dependant on the speed of the water entering the screw. The screw can be enclosed, but is in many cases left open.

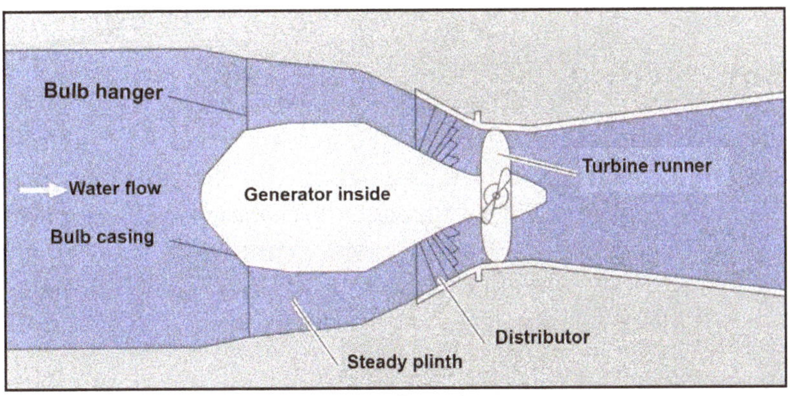

Figure: Bulb Turbine.

Power Developed by ROR Systems

There are two methods which can be used for calculating the power generated by run-of-river systems, the first based on the potential energy in stored water or head of water, and the second based on the kinetic energy of flowing water.

Potential Energy Method

Potential energy (PE) per unit volume.

$$\text{PE} = \rho g h$$

Where:

ρ = the density of the water (10^3 kg/m^3)

h = the head of water (m)

g = the gravitational constant (9,81 m/s^2)

The power P_{max} of the hydro system (in kW) is given by,

$$P_{max} = \eta \rho g h Q$$

Where:

Q is the volume of water flowing per second (the flow rate in m^3/s) η is the efficiency of the turbine,

For water flowing at 1 m^3/s from a head of 1 m, the power generated is equivalent to 9,81 kW assuming an energy conversion efficiency of 100% or just over 9 kW with a turbine efficiency of between 90% and 95%.

This simple looking formula usually applies to static hydro, using stored water. If applied to dynamic or kinetic systems, some attention needs to be given to the calculation of the values and how they influence the power output.

Kinetic Energy Method

The turbine converts the kinetic energy of the flowing water into the rotational (mechanical) energy of the turbine and the generator. The available energy therefore depends on the quantity of water flowing through the turbine and the square of its velocity. Impulse turbines which are only partially submerged are more commonly employed in fast flowing run-of-river installations while In deeper, slower flowing rivers, submerged Kaplan turbines may be used to extract the energy from the water flow.

The maximum power output from a turbine used in a kinetic ROR application is proportional to the kinetic energy of the water flowing through the turbine . Taking the efficiency η of the turbine into account, the maximum output power P_{max} is given by:

$$P_{max} = ½ \eta \rho Q v^2$$

Where:

v = the velocity of the water flow

Q = the volume of water flowing through the turbine per second

ρ = the density of water (10^3 kg/m^3)

but $Q = Av$, where A is the swept area of the turbine blades

Thus,

$$P_{max} = ½ \eta \rho A v^3$$

This relationship is directly analogous to the equation for the theoretical power generated by wind turbines. Note that the power output is proportional to the cube of the velocity of the water.

If we return to equation $P_{max} = \eta \rho g h Q$ and use the Bernoulli equation,

$$v = \sqrt{2gh} \ or \ v^2 \ 2gh$$

to get,

$$h = \frac{v2}{2g}$$

Substituting for h in equation $P_{max} = ½\eta \rho A v^3$.

Gives,

$$P_{max} = ½\eta \rho v^2 Q$$

Now $Q = vA$ so $P_{max} = ½\eta \rho A v^3$, the same result as in equation $P_{max} = ½\eta \rho A v^3$.

The result is:

- If the head or pressure drop is known but not the velocity, use equation $P_{max} = \eta \rho g h Q$.
- If the velocity is known neglect the head and use equation $P_{max} = ½\eta \rho A v^3$.

Thus the power generated by one cubic metre of water flowing at one metre per second through a turbine with 100% efficiency will be 0,5 kW or slightly less taking into account the inefficacies in the system. This is only one twentieth of the power generated by the same volume flow from the static system in the example above. To generate the same power with the same volume of water from a run-of-river installation, the speed of the water flow should be 3√20 metres per second (2,7 m/s).

Comparison to Traditional Hydro

There are several benefits that exist as a result of using run-of-the-river hydro instead of traditional, dam based hydro. First, traditional hydro dams are expensive and time consuming to build. Comparatively speaking, run-of-the-river systems are less expensive to build and can be built over a shorter period of time. In addition, many areas where large hydro is used frequently—such as in Canada—have developed many of the existing favourable hydropower sites.Run of river systems also avoid some of the environmental problems associated with the flooding, since the pondage is much smaller than the lakes for traditional hydro.

Although there are some favourable aspects to run-of-the-river hydro, the output is significantly lower than large scale hydro. Dam based hydroelectric generation usually provides a lower cost per kWh despite the larger initial investment. Therefore, the absence of the large dam and reservoir means that the power plant will be less reliable for electricity generation. If the water levels are depleted upstream, potentially due to drought conditions, there is less water to run the hydroelectric system.

Environmental Impact

Compared to the combustion of fossil fuels, run-of-the-river hydro is responsible for fewer greenhouse gas emissions. Most of these emissions are a result of the construction of the system itself, but the operation of the plant itself contributes almost no emissions. Run-of-the-river hydroelectric plants can also be beneficial compared to impoundment dams as the small amount of water storage results in a smaller environmental footprint when compared to dams with large amounts of water storage.

Although the emission impact of these systems is less than the combustion of fuels, there are other environmental impacts that must be taken into account. First, the manipulation of river flows can cause a significant number of environmental impacts. Any diversion in the river changes how the aquatic ecosystem works, which could affect fish populations and the health of the river overall. However, the small head difference in run-of-the-river systems allows for the construction of fish ladders, which could allow fish to move through the system without being harmed or having their migration routes interrupted. As well, the changes in the river basin or water composition could increase species mortality, disrupt migration, or cause an imbalance in biodiversity. Finally, thermal pollution and increased turbidity of the exiting water are possible side effects of sending water through turbines and back into a river.

Overall, it is difficult to determine in general whether or not the damage inflicted on the environment by run-of-the-river systems is outweighed by the relatively small output as compared to large hydro dams. That means that each project must be evaluated on the specific details of the hydroelectric power plants being proposed.

Advantages

When developed with care to footprint size and location, ROR hydro projects can create sustainable energy minimizing impacts to the surrounding environment and nearby communities. Advantages include:

Cleaner Power, Fewer Greenhouse Gases

Like all hydro-electric power, run-of-the-river hydro harnesses the natural potential energy of water, eliminating the need to burn coal or natural gas to generate the electricity needed by consumers and industry. Moreover, run-of-the-river hydro-electric plants do not have reservoirs thus eliminating the methane and carbon dioxide emissions caused by the decomposition of organic matter in the reservoir of a conventional hydro-electric dam. This is a particular advantage in tropical countries where methane generation can be a problem.

Less Flooding/Reservoirs

Without a reservoir, flooding of the upper part of the river does not take place. As a result, people remain living at or near the river and existing habitats are not flooded. Any pre-existing pattern of flooding will continue unaltered, presenting a flood risk to the facility and downstream areas.

Disadvantages

Unfirm Power

Run-of-the-River power is considered an "unfirm" source of power: a run-of-the-river project has little or no capacity for energy storage and hence can't co-ordinate the output of electricity generation to match consumer demand. It thus generates much more power during times when seasonal river flows are high (i.e., spring freshet), and depending on location, much less during drier summer months or frozen winter months.

Availability of Sites

Rapids can provide enough hydraulic head.

The potential power at a site is a result of the head and flow of water. By damming a river, the head is available to generate power at the face of the dam. Where a dam may create a reservoir hundreds of kilometres long, in run of the river the head is usually delivered by a canal, pipe or tunnel constructed upstream of the power house. Due to the cost of upstream construction, a steep drop is desirable, such as falls or rapids.

Environmental Impacts

Small, well-sited ROR projects can be developed with minimal environmental impacts. Larger projects have more environmental concerns. In the case of fish-bearing rivers a ladder may be required and dissolved gases downstream may affect fish.

In British Columbia the mountainous terrain and wealth of big rivers have made it a global testing ground for 10-50Mw run-of-river technology. As of March 2010, there were 628 applications pending for new water licences solely for the purposes of power generation – representing more than 750 potential points of river diversion.

Concerns

- Diverting large amounts of river water reduces river flows, affecting water velocity and depth, minimizing habitat quality for fish and aquatic organisms; reduced flows can lead to excessively warm water for salmon and other fish in summer.
- In undeveloped areas, new access roads and transmission lines can cause habitat fragmentation, allowing the introduction of invasive species.
- The lack of reservoir storage may result in intermittent operation, reducing the project's viability.

Pumped-storage Hydroelectricity

This method produces electricity to supply high peak demands by moving water between reservoirs at different elevations. At times of low electrical demand, the excess generation capacity is used to pump water into the higher reservoir. When the demand becomes greater, water is released back into the lower reservoir through a turbine. Pumped-storage schemes currently provide the most commercially important means of large-scale grid energy storage and improve the daily capacity factor of the generation system.

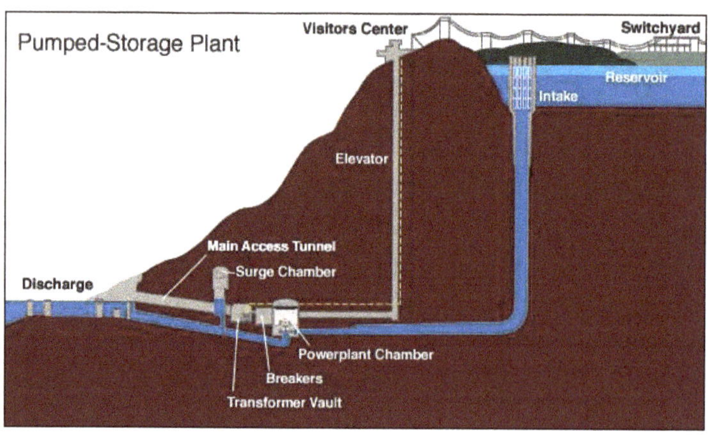

Figure: Diagram of the TVA pumped storage facility at Raccoon Mountain Pumped-Storage Plant.

Pumped storage is the largest-capacity form of grid energy storage available, The main disadvantage of PHS is the specialist nature of the site required, needing both geographical height and water availability. Suitable sites are therefore likely to be in hilly or mountainous regions, and potentially in areas of outstanding natural beauty, and therefore there are also social and ecological issues to overcome.

Figure: Pumped-storage hydroelectricity – the upper reservoir (Llyn Stwlan) and dam of the Ffestiniog Pumped Storage Scheme in north Wales. The lower power station has four water turbines which generate 360 MW of electricity within 60 seconds of the need arising.

At times of low electrical demand, excess generation capacity is used to pump water into the higher reservoir. When there is higher demand, water is released back into the lower reservoir through a turbine, generating electricity. Reversible turbine/generator assemblies act as pump and turbine (usually a Francis turbine design). Nearly all facilities use the height difference between two natural bodies of water or artificial reservoirs. Pure pumped-storage plants just shift the water between reservoirs, while the "pump-back" approach is a combination of pumped storage and conventional hydroelectric plants that use natural stream-flow. Plants that do not use pumped-storage are referred to as conventional hydroelectric plants; conventional hydroelectric plants that have significant storage capacity may be able to play a similar role in the electrical grid as pumped storage, by deferring output until needed.

Taking into account evaporation losses from the exposed water surface and conversion losses, energy recovery of 80% or more can be regained. The technique is currently the most cost-effective means of storing large amounts of electrical energy on an operating basis, but capital costs and the presence of appropriate geography are critical decision factors.

The relatively low energy density of pumped storage systems requires either a very large body of water or a large variation in height. For example, 1000 kilograms of water (1 cubic meter) at the top of a 100 meter tower has a potential energy of about 0.272 kW·h (capable of raising the temperature of the same amount of water by only 0.23 Celsius = 0.42 Fahrenheit). The only way to store a significant amount of energy is by having a large body of water located on a hill relatively near, but as high as possible above, a second body of water. In some places this occurs naturally, in others one or both bodies of water have been man-made. Projects in which both reservoirs are artificial and in which no natural waterways are involved are commonly referred to as "closed loop".

This system may be economical because it flattens out load variations on the power grid, permitting thermal power stations such as coal-fired plants and nuclear power plants that provide base-load electricity to continue operating at peak efficiency (Base load power plants), while reducing the need for "peaking" power plants that use the same fuels as many base load thermal plants, gas and oil, but have been designed for flexibility rather than maximal thermal efficiency. However, capital costs for purpose-built hydro-storage are relatively high.

Along with energy management, pumped storage systems help control electrical network frequency and provide reserve generation. Thermal plants are much less able to respond to sudden changes in electrical demand, potentially causing frequency and voltage instability. Pumped storage plants, like other hydroelectric plants, can respond to load changes within seconds.

The important use for pumped storage is to level the fluctuating output of intermittent energy sources. The pumped storage provides a load at times of high electricity output and low electricity demand, enabling additional system peak capacity. In certain jurisdictions, electricity prices may be close to zero or occasionally negative (Ontario in early September, 2006), on occasions that there is more electrical generation than load available to absorb it; although at present this is rarely due to wind alone, increased wind generation may increase the likelihood of such occurrences. It is particularly likely that pumped storage will become especially important as a balance for very large scale photovoltaic generation.

Potential Technologies

Seawater

Pumped storage plants can operate with seawater, although there are additional challenges compared to using fresh water. In 1999, the 30 MW Yanbaru project in Okinawa was the first demonstration of seawater pumped storage. It has since been decommissioned. A 300 MW seawater-based Lanai Pumped Storage Project was considered for Lanai, Hawaii, and seawater-based projects have been proposed in Ireland. A pair of proposed projects in the Atacama Desert in northern Chile would use 600 MW of photovoltaic solar (Skies of Tarapacá) together with 300 MW of pumped storage (Mirror of Tarapacá) raising seawater 600 metres (2,000 ft) up a coastal cliff.

Underground Reservoirs

The use of underground reservoirs has been investigated. Recent examples include the proposed Summit project in Norton, Ohio, the proposed Maysville project in Kentucky (underground limestone mine), and the Mount Hope project in New Jersey, which was to have used a former iron mine as the lower reservoir. The proposed energy storage at the Callio site in Pyhäjärvi (Finland) would utilize the deepest base metal mine in Europe, with 1,450 metres (4,760 ft) elevation difference. Several new underground pumped storage projects have been proposed. Cost-per-kilowatt estimates for these projects can be lower than for surface projects if they use existing underground mine space. There are limited opportunities involving suitable underground space, but the number of underground pumped storage opportunities may increase if abandoned coal mines prove suitable.

In Bendigo, Victoria, Australia, the Bendigo Sustainability Group has proposed the use of the old gold mines under Bendigo for Pumped Hydro Energy Storage. Bendigo has the greatest concentration of deep shaft hard rock mines anywhere in the world with over 5,000 shafts sunk under Bendigo in the second half of the 19th Century. The deepest shaft extends 1,406 metres vertically underground. A recent pre-feasibility study has shown the concept to be viable with a generation capacity of 30 MW and a run time of 6 hours using a water head of over 750 metres.

Decentralised Systems

Small pumped-storage hydropower plants can be built on streams and within infrastructures, such as drinking water networks and artificial snow making infrastructures. Such plants provide distributed energy storage and distributed flexible electricity production and can contribute to the decentralized integration of intermittent renewable energy technologies, such as wind power and solar power. Reservoirs that can be used for small pumped-storage hydropower plants could include natural or artificial lakes, reservoirs within other structures such as irrigation, or unused portions of mines or underground military installations. In Switzerland one study suggested that the total installed capacity of small pumped-storage hydropower plants in 2011 could be increased by 3 to 9 times by providing adequate policy instruments.

Underwater Reservoirs

In March 2017 the research project StEnSea (Storing Energy at Sea) announced their successful completion of a four-week test of a pumped storage underwater reservoir. In this configuration a hollow sphere submerged and anchored at great depth acts as the lower reservoir, while the upper reservoir is the enclosing body of water. Electricity is created when water is let in via a reversible turbine integrated into the sphere. During off-peak hours the turbine changes direction and pumps the water out again, using “surplus” electricity from the grid. The quantity of power created when water is let in grows proportionally to the height of the column of water above the sphere, in other words: the deeper the sphere is located, the more potential energy it can store, which can be transformed into electric power. On the other hand, pumping the water back out at greater depths also uses up more power, since the turbine-turned-pump must act on the same entire column of water.

As such the energy storage capacity of the submerged reservoir is not governed by the gravitational energy in the traditional sense, but rather by the vertical pressure variation.

While StEnSea’s test took place at a depth of 100 m in the fresh water Lake Constance, the technology is foreseen to be used in salt water at greater depths. Since the submerged reservoir needs only a connecting electrical cable, the depth at which it can be employed is limited only by the depth at which the turbine can function, currently limited to 700 m. The challenge of designing salt water pumped storage in this underwater configuration brings a range of advantages:

- No land area is required,
- No mechanical structure other than the electrical cable needs to span the distance of the potential energy difference,
- In the presence of sufficient seabed area multiple reservoirs can scale the storage capacity without limits,
- Should a reservoir collapse, the consequences would be limited apart from the loss of the reservoir itself,
- Evaporation from the upper reservoir has no effect on the energy conversion efficiency,
- Transmission of electricity between the reservoir and the grid can be established from a nearby offshore wind farm limiting transmission loss and obviating the need for onshore cabling permits.

A current commercial design featuring a sphere with an inner diameter of 30 m submerged to 700 m would correspond to a 20 MWh capacity which with a 5 MW turbine would lead to a 4-hour discharge time. An energy park with multiple such reservoirs would bring the storage cost to around a few eurocents per kWh with construction and equipment costs in the range €1,200-€1,400 per kW. To avoid excessive transmission cost and loss, the reservoirs should be placed off deep water coasts of densely populated areas, such as Norway, Spain, USA and Japan. With this limitation the concept would allow for worldwide electricity storage of close to 900 GWh.

For comparison, a traditional, gravity-based pumped storage capable of storing 20 MWh in a water reservoir the size of a 30 m sphere would need a hydraulic head of 519 m with the elevation spanned by a pressurized water pipe requiring typically a hill or mountain for support.

Home use

Using a pumped-storage system of cisterns and small generators, pico hydro may also be effective for “closed loop” home energy generation systems.

Pico Hydropower

Pico hydro is hydro power with a maximum electrical output of five kilowatts (5kW). Hydro power systems of this size benefit in terms of cost and simplicity from different approaches in the design, planning and installation than those which are applied to larger hydro power. Recent innovations in Pico-hydro technology have made it an economic source of power even in many of the world's poorest and most inaccessible areas. It is also a versatile power source. AC electricity can be produced enabling standard electrical appliances to be used. Common examples of devices which can be powered by Pico-hydro are light bulbs, radio and televisions. Normally, Pico-hydropower system is found at rural or hilly area. Figure below shows an example of typical Pico hydro system applications at hilly area. This system will operate using upper water reservoir which is a few meter high from ground. From the reservoir, water flows downhill through the piping system. This downhill distance is called "head" and it allows the water to accelerate for prime moving system. Thus, the turbine will rotate the alternator to produce electricity. However, this research is conducted to show the potential of consuming water distributed to houses at town area as an alternative of renewable energy source.

The water flow inside the pipelines has potential of kinetic energy to spin small scale generator turbine for electricity generation. Therefore, this project has been done to show the additional use of consuming water distributed to houses for electrical power generation instead of routine activities such as bathe, laundry and dish wash. The electricity can be generated at the same time those usual activities are done without extra charge on the water bill consumption. The main function of the system is to store the generated power by means of battery charging for future use particularly during electricity blackouts. The proposed system is expected has a maximum capacity of 10W which is very much less compared to other Pico-hydro power systems.

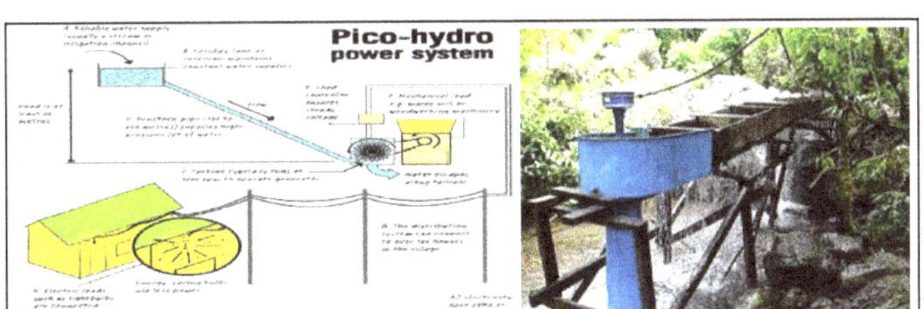

Figure: Pico-hydro power system applications at rural area.

Pico-hydro System Planning & Development

There are many factors that determine the feasibility and achievability of the system. This includes:

i. The amount of power available from the water flow inside the pipelines. This depends on the water pressure, amount of water available and friction losses in the pipelines.

ii. The turbine type and availability of required generator type and capacity.

iii. The types and capacity of electrical loads to be supplied by the Pico-hydro system.

iv. The cost of developing the project and operating the system.

(A) Power Estimation: In general, the feasibility of the proposed Pico-hydro system is based on the following potential input and output power equation:

$$Pin = H \times Q \times g \quad (1)$$

$$Pout = H \times Q \times g \times \eta \quad (2)$$

Where,

Pin = Input power (Hydro power)

Pout =Output power (Generator output)

H = Head (meter)

Q = Water flow rate (liter/second)

g = gravity (9.81 m/s2)

η = efficiency

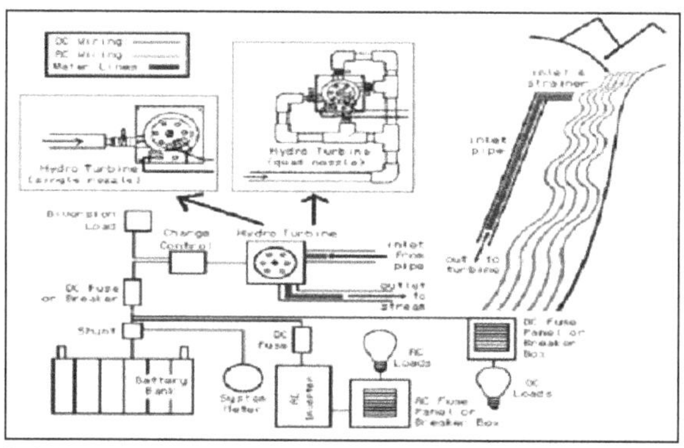

Figure: Probable Pico-hydro system: Sustainable Renewable Energy Solution.

Based on the equation (Pin = H x Q x g (1)) and (Pout= H x Q x g x η (2)), both head and water flow rate are very important parameters in hydropower system. Head is a measure of falling water at turbine, i.e. vertical distance from the top of the penstock to the turbine at the bottom. Conversely, water flow rate is the amount of water flows within one second. Normally, water flow available is more than needed since the flows for Pico-hydro are small. Thus, it is important to measure the head carefully because the greater head, the greater power and the higher speed of the turbine rotation.

(B) Head Measurement: When determining head (falling water), gross or "static" head and net or "dynamic" head must be considered. Gross head is the vertical distance between the top of the penstock and the point where the water hits the turbine. Net head is gross head minus the pressure or head losses due to friction and turbulence in the penstock. These head losses depend on the type, diameter, and length of the penstock piping, and the number of bends or elbows. Gross head can be used to estimate power availability and determine general feasibility, but net head is used to calculate the actual power available. There are many methods of head measurement. However, since the proposed Pico-hydro system uses consuming water distributed to houses supplied by the Water Utility Company whereby the utility water tank can be very far from the houses, the simplest and most practical method for head measurement is water-filled tube and calibrated pressure gauge.

Through this method, the pressure gauge reading in psi can be converted to head in meters using the following equation of pressure to head conversion:

$$H = 0.704 \times P \quad (3)$$

Where = Head (meter) & P = Pressure (psi)

Equation above shows that the water pressure at consumers" end is a very important parameter to be determined in the design and development of the proposed Pico-hydro system. The water pressure represents the net head of the system that useful to calculate the actual power available.

(C) Water Flow Rate Measurement: The most simple of flow measurement for small streams is the bucket method. Therefore, this method has been used due to the capacity of the proposed hydropower system is significantly small. Moreover, this method is considerably practical due to the proposed hydropower system is very uncommon compared to other system in its category in which the source of energy is from the consuming water distributed to houses by the Water Utility Company. Throughout this method, the flow rate of the distributed water is diverted into a bucket or barrel and the time it takes for the container to fill is recorded. The volume of the container is known and the flow rate is simply obtained by dividing this volume by the filling time. For example, the flow rate of water that filled 20 litres bucket within one minute is 20 litres per minute or 0.333 l/s. This is repeated several times to give more consistent and accurate measurement.

(D) Pipeline System and Friction Loss: Piping system is used to carry water to a turbine. This is commonly termed as penstock which consists of pipe from the reservoir or fore bay tank to the turbine and valve or gate that controls the rate of water flow. The proposed Pico-hydro system will have the water source from consuming water distributed to houses. Thus, the system must be designed with ability to produce high water pressure to rotate the turbine at the most possible speed and at the same time the water can be recycled and used to other routine activities such as bathe and laundry. Thus, no extra charge on the water bill consumption incurred. In order to do so, a suitable piping scheme with appropriate nozzle between the source (consumers" end) and the turbine is required to maximize the turbine rotation speed.

Figure: Pen stroke for Pico-hydro system.

In real fluid flows, friction losses occur due to the resistance of the pipe walls and the fittings. This leads to an irreversible transformation of the flowing fluid energy into heat. Friction head loss is divided into two main categories, "major losses" associated with energy loss per length of pipe, and "minor losses" associated with bends, fittings, valves, etc. For the proposed system, the hydro power available at the consumers‟ water outlet (consumers‟ end) is the net power after taking into account the friction losses along the pipelines from the utility tank to the consumers. Thus, consideration on the pipe length and diameter to handle the amount of water flow and piping accessories to convey the water to the turbine is very important to minimize the friction loss for the piping scheme between the source and the turbine of the Pico-hydro power system. This can be done by appropriately select the diameters and types of bends, fittings and valve and minimizing the use of these accessories. Moreover, it is necessary to minimize the piping system length between the water source and the turbine although it is extremely short when compared to the main pipeline from the utility tank to consumers‟ end. By considering all these matters, the proposed piping Pico-hydro piping scheme is assumed to have minor friction losses or can be neglected. This means the net hydro power at consumers‟ end is more or less similar to hydro power to turbine. As a result, the following are the list of piping accessories that used in the system:

i. Valve – ball valve

ii. Nozzle – variable

iii. Elbow – 90 degree

iv. Tee – flanged

v. Straight connector

vi. Pressure Gauge – 0 to 10 bar

vii. Main pipe – diameter 1.5mm

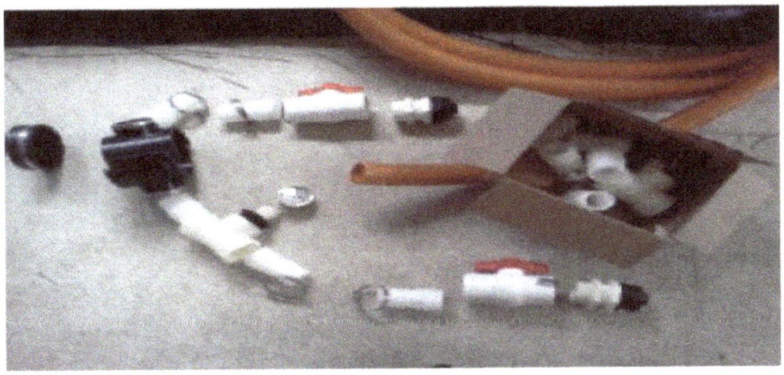

Figure: Pico hydro Piping Control elements.

(E) Selection of Generator: Generating system for a hydro power scheme is selected based on the following concerns:

i. The estimated power of a hydropower system.

ii. Type of supply system and electrical load: AC or DC.

iii. Available generating capacity in the market.

iv. Generator with cost effective.

Normally, Pico-hydro systems use AC generator either induction or synchronous machine type. This is because the system is used to supply AC electrical appliances and DC generator with size above 2kW is said expensive and has brush gear that requires appreciable maintenance. In addition, DC switches for the voltages and currents concerned are more expensive than their AC equivalents. However, in this project, a brush permanent magnet DC Generator is preferred as the main function of the proposed Pico-hydro system is for energy storage (battery charging). One significant advantage of using DC type of permanent magnet generator over AC generator is that DC generator is designed to provide high currents at minimum voltage requirement for the charging of battery and operation of direct current loads.

This is related with the load type to be supplied. Moreover, permanent magnet generator is selected as it is much cheaper and has smaller overall size rather than of wound field. Other than that, this type of generator is more efficient because no power is wasted to generate the magnetic field. Hence, permanent magnet DC generator manufactured by "Wind stream Power" with maximum power output of 80W at 2800rpm shaft rotation is chosen. The capacity of the generator is considerably high compared to the estimated power mentioned earlier, i.e. 10W. This is because it is very difficult to find the generating capacity available in the market that match with the estimated power. The maximum current at continuous duty is 1.5A whilst maximum current for 10minutes duty is limited to 2.5A. One indisputable problem that will be encountered when using this generator is high torque during the shaft.

Figure: Pico hydro Generator rotation.

This makes the rotation of the generator is not at the highest possible speed. Research on the generator design itself to maximize the system efficiency will be done in the future.

(F) Selection of Turbine Type: Selection of turbine to be used is very important in the design and development of a hydropower system. Table shows the groups of impulse and reaction turbines that are available. In general, reaction turbine is fully immersed in water and is enclosed in a pressure casing. The runner or rotating element and casing are carefully engineered so that the clearance between them is minimized. In contrast, impulse turbine can operate in air and works with high-speed jet of water.

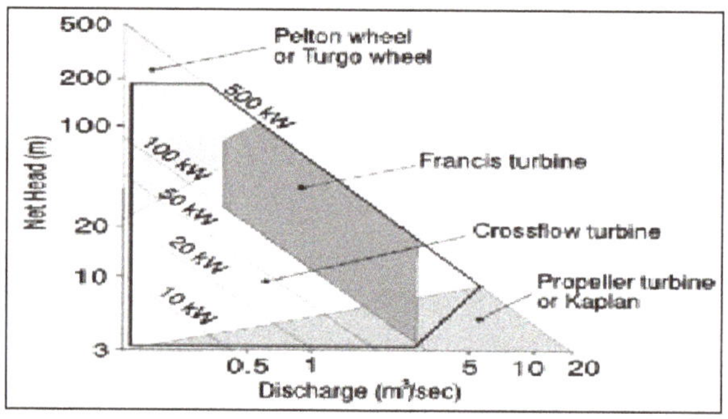

Figure: Head-flow ranges of small hydro turbines.

Usually, impulse turbines are cheaper than reaction turbines because no specialist pressure casing and no carefully engineered clearance are needed.

Pelton Turbine: Pelton turbines are suitable to high head, low flow applications. It can be outfitted with one, two, or more nozzles for higher output. Typically, when using this type of turbine, water is piped down a hillside so that at the lower end of the pipe, it emerges from a narrow nozzle as a jet with very high velocity to the turbine blades.

The similar concept of water jet is also utilized in the proposed Pico-hydro system; instead the water is tapped from the domestic consuming water outlet. After a series of performance test on various constructions of Pelton turbine blades, the turbine with 6 cups fabricated from polyvinyl chloride (PVC) as shown in figure below is used for the proposed system.

The advantages of using PVC are less expensive and not require specialist to fabricate the turbine blades. Meanwhile, the base that associated with the drive shaft (rotor) for the turbine blades mounting is fabricated from aluminum with 80mm diameters. Small size of turbine is important to ensure the generator shaft to rotate at an optimum speed.

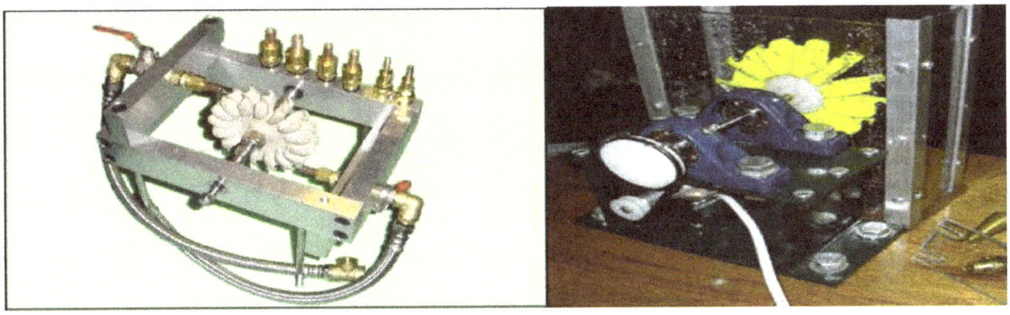

Figure: Turbine for Pico Hydro System.

Table: Groups of Impulse and Reaction Turbine.

Turbine Runner	Head Pressure		
	High	Medium	Low
Impulse	Pelton Turgo Multi-jet Pelton	Cross flow Turgo Multi-jet Pelton	Cross flow
Reaction		Francis Pump-as-turbine	Propeller Kaplan

(G) Types of Electrical Loads: Electrical loads that are normally connected to a Pico-hydro system at rural area are lighting, battery chargers, radios, televisions, ventilation fans and refrigerators. For the proposed Pico-hydro system, however, the generating capacity is much lower compared to the existing Pico-hydro system at rural area. Thus, the main function of the proposed system is for battery charging. A battery allows the future use of small electrical loads and can be recharged when required. Examples of future use of small loads particularly during electricity blackouts are LED lighting, mobile phone battery charging and toys battery charging. There are two common types of rechargeable batteries used for providing power to small loads which are lead-acid and nickel-cadmium (Ni-Cad). Battery size depends on the generating capacity of the proposed Pico-hydro system and its application. For this type of Pico-hydro system, Ni-Cad battery is preferred and more practical as it is easier to handle and reliable. Moreover, Ni-Cad is the opposite to lead acid in that it performs better and last longer if fully discharged before re-charging. On the other hand, lead acid batteries should not be fully discharged as this damages them.

(H) Battery Charger: As illustrated in figure below, the generator output is connected to the charging circuits for energy storage purpose. Simple charging circuit as shown in figure below is used. The battery charger is suitable for 9V to 12V batteries. For charging purpose, the maximum load current is limited to 1.5A. This is based on the maximum load current of the LM317 voltage regulator and the maximum current at continuous duty of the generator.

Figure: Pico hydro in operation.

In addition, due to the generating capacity of the Pico-hydro system, Ni-Cad battery is preferred. Figure above also illustrates that there is a point of direct output from generator. This is to offer other application that might be relevant.

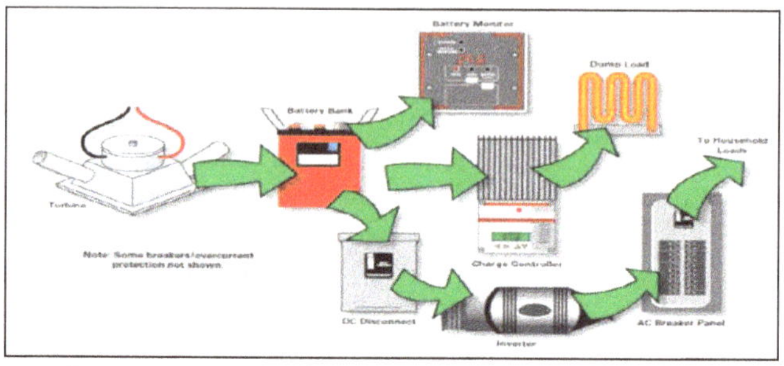

Figure: Operation of Pico hydro with modern controls.

Micro Hydropower

Micro hydropower is the small-scale harnessing of energy from falling water, such as steep mountain rivers. Using this renewable, indigenous, non-polluting resource, micro-hydro plants can generate power for homes, hospitals, schools and workshops. The micro hydropower station, which converts the energy of flowing water into electricity, provides poor communities in rural areas with an affordable, easy to maintain and long-term solution to their energy needs.

A micro hydro power (MHP)'plant' is a type of hydro electric power scheme that produces up to 100 KW of electricity using a flowing steam or a water flow. The electricity from such systems is used to power up isolated homes or communities and is sometimes connected to the public grid.

Micro hydro systems are generally used in developing countries to provide electricity to isolated communities or rural villages where electricity grid is not available. Feeding back into the national grid when electricity production is in surplus is also evident in some cases. The micro hydro scheme design can be approached as per household basis or at the village level often involving local materials and labor.

In 1995, the micro-hydro capacity in the world was estimated at 28 GW, supplying about 115 TWh of electricity. About 60% of this capacity was in the developed world, with 40% in developing areas.

Micro hydro plants that are found in the developing world are mostly in mountainous regions for instance in the some places in the Himalayas as well as in Nepal where there are around 2,000 schemes, including both mechanical and electrical power generation. In South America, there are micro-hydro programs in the countries along the Andes, such as Peru and Bolivia. Smaller programs have also been set up in the hilly areas of Sri Lanka, Philippines and some parts of China.

Technology

Scheme Components

Most micro-hydro systems are 'run-of-river' which means that they don't need large dams to store water. However, they do need some water-management systems.

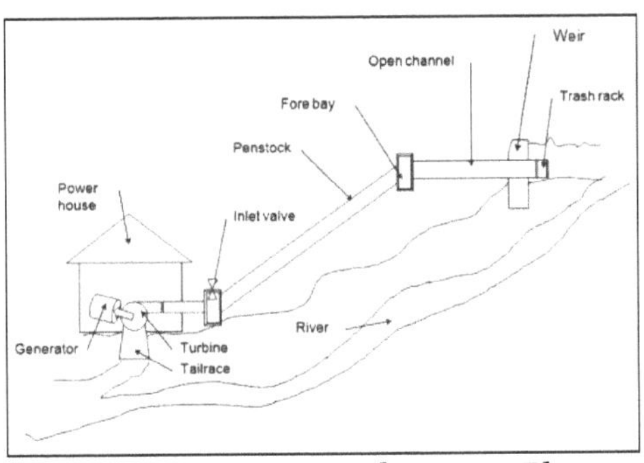

Components Micro Hydro Power Plant.

The illustration above shows just how a micro-hydro system can be setup. For water diversion the river water level has to be raised by a barrier, the weir. The water is diverted at the intake and conveyed by the channel along the landscape´s contour lines. The spillways protect against damage from excessive water flow. Water is slowed down and collected in the fore bay, from where it enters into the penstock; the pressure pipe conveys the water to the power house where the power conversion turbine, mill or generating equipment is installed. The turbine is the core of a MHP, which is rotated by the moving water. Different types of turbines are used depending on the head and the flow of the site the turbines are used to rotate a shaft which is then used to drive the generator. The water is then discharged via the draft tube or a tail race channel in case of cross flow or Pelton turbines.

Due to the nature of the micro-hydro schemed to be remote; a local grid is constructed to distribute the electricity to the different users. The demand output must match the capacity of the generator otherwise the voltage and frequency can vary suddenly, which can result in the damage of certain electrical equipment. The power demand in an off-grid is often variable since people switch lights and machines on and off, so the supply from the micro-hydro system must be varied to keep close control. This can be done by varying the water flow, or by using an electronic load controller.

Suitable Conditions for Micro Hydro Power Plants

The ideal geographical areas for exploiting small scale hydro schemes is where there are steep rivers flowing all year round. Islands with moist marine climates are also suitable. Low-head turbines have been developed for small-scale exploitation of rivers or irrigation canals where there is a small head but sufficient flow to provide adequate power.

To understand more about a suitable potential site, the hydrology of the site needs to be known and a site survey carried out so as to determine the actual flow and head data. Hydrological information is easily accessible from the metrological or irrigation department of the particular national government. Site surveys usually give a more detailed information of the site conditions to allow power calculation to be done and design work to begin. Flow data should however be collected over a period of one year where possible, this is to ascertain on the fluctuation in the river flow over the various seasons.

Micro Hydro Power Plant Development - Barriers

There are various barriers that hinder the dissemination of MHP, some of them have been identified as:

- Policy and regulatory framework: In most cases there exists no sufficient policies and frameworks that govern MHP schemes, this is because the MHP is either not regulated at all or is combined with a broader framework made for rural electrification which may be unclear and in transparent. Such challenges causes the MHP project developers not to know which requirements apply and work in an unreliable grey area of regulation.
- Financing: Lack of sufficient funding to be used in development is a common challenge as most MHP rely on donor funding which in most cases is only available in funding a small portion of the hydro power potential. One of the ways that such as case can be addressed is if there can be an option of exploring other sources of funding especially from private venture capitalists and local banks.
- Capacity to plan, build and operate MHP plants: Lack of knowledge and awareness on MHP potential posses a great challenge for rural electrification, hydro power schemes still dominate as political decision makers still tend to go for them as a more "modern" approach. Combined with that there is minimal

capacity to design, implement and revise the MHP supportive policies and regulations. And at the technical level, local capacity is often missing to plan, build and run MHP projects. The is also a problem in the lack of a ready supply of affordable turbine parts and the lack of domestic manufacturing capacity for hydro systems of all sizes also poses a barrier to a swift and cost-effective MHP project development.

- Data on hydro resources: There is usually a lack of interest in MHP deployment from the politicians and power utilities companied by the lack appropriate capacities and budgets, as well as unavailability of pubic data on MHP sites. Such a lack of sound basic data (e.g. on mid-to long-term hydrological, geographic, geologic data and figures on the current and future demand for electricity and social infrastructure, but especially on effects of seasonal and long-term river flow variations), poses a major barrier for private investors in MHP. This causes a bottleneck for investment in hydropower systems as there is an increase in climate variability accompanied with the destruction of rainfall catchment areas.

Application: Use of Micro Hydro Power Plants

Power produced from a small hydro station can be used for various purposes, some of the uses have been classified as follows:

1. Productive Use: This is where the electricity generated is used to perform activities where money is exchanged for a service. Most of this scenarios take place in small businesses.
2. Consumptive Use: All the other used that the electricity can be used for are called consumptive use. They include using the electricity at the household or close to the household.

Besides the productive and consumptive use, a distinction can also be made between the use of power in a mechanical way or in the form of electricity:

	mechanic	electricity
productive use	▪ agro processing ▪ timber sawing ▪ textile fabrication ▪ cooling ▪ drying	▪ mechanical uses with electricity as intermediate ▪ heating ▪ lighting ▪ fertiliser production
consumptive use		▪ domestic lighting ▪ cooking ▪ cooling ▪ radio and television

As the above illustration shows power that is generated by MPH is a convenient source of electricity to fuel anything from workshop machines to domestic lighting as the power can also be supplied to villages via portable rechargeable batteries and thus there are no expensive connection costs. Batteries can as well be charged and used to provide the local community with power. For industrial use however, the turbine shaft can be used directly as mechanical power as opposed to converting it into electricity via generator or batteries. This is suitable for agro-processing activities such as milling, oil extraction and carpentry.

Micro Hydro Pros – Advantages

Efficient Energy Source

It only takes a small amount of flow (as little as two gallons per minute) or a drop as low as two feet to generate electricity with micro hydro. Electricity can be delivered as far as a mile away to the location where it is being used.

Reliable Electricity Source

Hydro produces a continuous supply of electrical energy in comparison to other small-scale renewable technologies. The peak energy season is during the winter months when large quantities of electricity are required.

No Reservoir Required

Microhydro is considered to function as a 'run-of-river' system, meaning that the water passing through the generator is directed back into the stream with relatively little impact on the surrounding ecology.

Cost Effective Energy Solution

Building a small-scale hydro-power system can cost from $1,000 – $20,000, depending on site electricity requirements and location. Maintenance fees are relatively small in comparison to other technologies.

Power for Developing Countries

Because of the low-cost versatility and longevity of micro hydro, developing countries can manufacture and implement the technology to help supply much needed electricity to small communities and villages.

Integrate with the Local Power Grid

If your site produces a large amount of excess energy, some power companies will buy back your electricity overflow. You also have the ability to supplement your level of micro power with intake from the power grid.

Micro Hydro Cons – Disadvantages

Suitable Site Characteristics Required

In order to take full advantage of the electrical potential of small streams, a suitable site is needed. Factors to consider are: distance from the power source to the location where energy is required, stream size (including flow rate, output and drop), and a balance of system components — inverter, batteries, controller, transmission line and pipelines.

Energy Expansion not Possible

The size and flow of small streams may restrict future site expansion as the power demand increases.

Low-power in the Summer Months

In many locations stream size will fluctuate seasonally. During the summer months there will likely be less flow and therefore less power output. Advanced planning and research will be needed to ensure adequate energy requirements are met.

Environmental Impact

The ecological impact of small-scale hydro is minimal; however the low-level environmental effects must be taken into consideration before construction begins. Stream water will be diverted away from a portion of the stream, and proper caution must be exercised to ensure there will be no damaging impact on the local ecology or civil infrastructure.

Small Hydro

Small scale hydropower systems capture the energy in flowing water and convert it to usable energy. Although the potential for small hydro-electric systems depends on the availability of suitable water flow, where the resource exists it can provide cheap clean reliable electricity. A well designed small hydropower system can blend with its surroundings and have minimal negative environmental impacts.

Moreover, small hydropower has a huge, as yet untapped potential in most areas of the world and can make a significant contribution to future energy needs. It depends largely on already proven and developed technology, yet there is considerable scope for development and optimization of this technology.

Working of Small Hydro

Hydropower systems use the energy in flowing water to produce electricity or mechanical energy. The water flows via channel or penstock to a waterwheel or turbine where it strikes the bucket of the wheel, causing the shaft o f the waterwheel or turbine to rotate. When generating electricity, the rotating shaft, which is connected to an alternator or generator, converts the motion of the shaft into electrical energy. This electrical energy may be used directly, stored in batteries, or inverted to produce utility-quality electricity. A small scale hydroelectric facility requires that a sizable flow of water and a proper height of fall of water, called head, is obtained without building elaborate and expensive facilities. Small hydroelectric plants can be developed at existing dams and have been constructed in connection with river and lake water-level control and irrigation schemes. By using existing structures, only minor new civil engineering works are required, which reduces the cost of this component of a development.

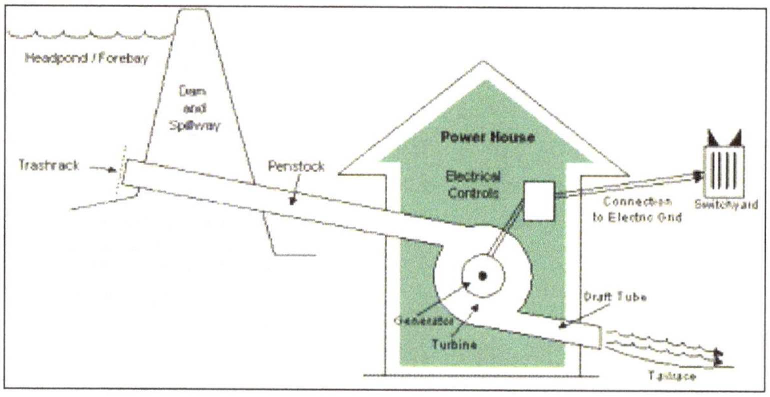

Parts of a Small Scale Hydropower Facility

In other, more rugged regions of the country, it is possible to develop relatively higher heads without elaborate or expensive civil engineering works so that relatively smaller flows are required to develop the desired power. In these cases, it may be possible to construct a relatively simple diversion structure and obtain the highest drop by diverting flows at the top of a waterfall or steeply falling watercourse.

Project Design

Many companies offer standardized turbine generator packages in the approximate size range of 200 kW to 10 MW. These "water to wire" packages simplify the planning and development of the site since one vendor looks after most of the equipment supply. Because non-recurring engineering costs are minimized and development cost is spread over multiple units, the cost of such package systems is reduced. While synchronous generators capable of isolated plant operation are often used, small hydro plants connected to an electrical grid system can use economical induction generators to further reduce installation cost and simplify control and operation.

Small "run of the river" projects do not have a conventional dam with a reservoir, only a weir to form a headpond for diversion of inlet water to the turbine. Unused water simply flows over the weir and the headpond may only be capable of a single day's storage, not enough for dry summers or frozen winters when generation may come to a halt. A preferred scenario is to have the inlet in an existing lake.

Countries like India and China have policies in favor of small hydro, and the regulatory process allows for building dams and reservoirs. In North America and Europe the regulatory process is too long and expensive to consider having a dam and a reservoir for a small project.

Small hydro projects usually have faster environmental and licensing procedures, and since the equipment is usually in serial production, standardized and simplified, and the civil works construction is also reduced, the projects may be developed very rapidly. The physically smaller size of equipment makes it easier to transport to remote areas without good road or rail access.

One measure of decreased environmental impact with lakes and reservoirs depends on the balance between stream flow and power production. Reducing water diversions helps the river's ecosystem, but reduces the hydro system's return on Investment (ROI). The hydro system design must strike a balance to maintain both the health of the stream and the economics.

Benefits of Small-scale Hydropower

Hydroelectric systems provide the following general benefits:

- Hydroelectric energy is a continuously renewable electrical energy source.
- Hydroelectric energy is non-polluting - no heat or noxious gases are released.
- Hydroelectric energy has no fuel cost and with low operating and maintenance costs, it is essentially inflation proof.
- Hydroelectric energy technology is a proven technology that offers reliable and flexible operation.
- Hydroelectric stations have a long life and many existing stations have been in operation for more than half a century and are still operating efficiently.
- Hydropower station efficiencies of over 90% are achieved making it the most efficient of energy conversion technologies.

References

- Conduit-hydropower-recovering-energy-one-control-valve-at-a-time-5519: valvemagazine.com, Retrieved 11, March 2020
- Hydroelectricity: newworldencyclopedia.org, Retrieved 13, May 2020
- Hydropower-and-pumped-storage-22104: altenergymag.com, Retrieved 24, June 2020

- Tidalenergy: conserve-energy-future.com, Retrieved 20, February 2020
- Pelc, Robin; Fujita, Rod M. (November 2002). "Renewable energy from the ocean". Marine Policy. Elsevier. 26 (6): 471–479. doi:10.1016/S0308-597X(02)00045-3
- Hydropower-learning-centre/head-and-flow-detailed-review: renewablesfirst.co.uk, Retrieved 26, July 2020

Chapter 4

Watermills and Water Turbines

Watermill is a mill that uses a water turbine to operate a mechanical process. A water turbine is a rotary machine that converts the kinetic and potential energy of water into mechanical or electrical energy. All the diverse aspects of watermills and water turbines have been carefully analyzed in this chapter.

Watermill

A watermill or water mill is a mill that uses moving water as its power source. It is thus a structure that uses a water wheel or water turbine to drive a mechanical process such as milling (grinding), rolling, or hammering. Such processes are needed in the production of many material goods, including flour, lumber, paper, textiles, and many metal products. Thus watermills may be gristmills, sawmills, paper mills, textile mills, hammermills, trip hammering mills, rolling mills, wire drawing mills, and so on.

Interior of the Lyme Regis watermill, UK (14th century).

One major way to classify watermills is by wheel orientation (vertical or horizontal), one powered by a vertical waterwheel through a gearing mechanism, and the other equipped with a horizontal waterwheel without such a mechanism. The former type can be further divided, depending on where the water hits the wheel paddles, into undershot, overshot, breastshot and pitchback (backshot or reverse shot) waterwheel mills. Another way to classify water mills is by an essential trait about their location: tide mills use the movement of the tide; ship mills are water mills onboard (and constituting) a ship.

History

Classical Antiquity

Model of a Roman type Roman water-powered grain-mill described by Vitruvius. The millstone (upper floor) is powered by an undershot waterwheel by the way of a gear mechanism (lower floor).

Vertical axle watermill in Dalarna, Sweden.

Hellenistic engineers invented the two main components of watermills, the waterwheel and toothed gearing, and were, along with the Romans, the first to operate undershot, overshot and breastshot waterwheel mills.

The earliest evidence of a water-driven wheel is probably the Perachora wheel (3rd century BC), in Greece. The earliest written reference is in the technical treatises *Pneumatica* and *Parasceuastica* of the Greek engineer Philo of Byzantium (c. 280–220 BC). The British historian of technology M.J.T. Lewis has shown that those portions of Philo of Byzantium's mechanical treatise which describe water wheels and which have been previously regarded as later Arabic interpolations, actually date back to the Greek 3rd-century BC original. The sakia gear is, already fully developed, for the first time attested in a 2nd-century BC Hellenistic wall painting in Ptolemaic Egypt.

Lewis assigns the date of the invention of the horizontal-wheeled mill to the Greek colony of Byzantium in the first half of the 3rd century BC, and that of the vertical-wheeled mill to Ptolemaic Alexandria around 240 BC.

The Greek geographer Strabon reports in his *Geography* a water-powered grain-mill to have existed near the palace of king Mithradates VI Eupator at Cabira, Asia Minor, before 71 BC.

The Roman engineer Vitruvius has the first technical description of a watermill, dated to 40/10 BC; the device is fitted with an undershot wheel and power is transmitted via a gearing mechanism. He also seems to indicate the existence of water-powered kneading machines.

The Greek epigrammatist Antipater of Thessalonica tells of an advanced overshot wheel mill around 20 BC/10 AD. He praised for its use in grinding grain and the reduction of human labour:

Hold back your hand from the mill, you grinding girls; even if the cockcrow heralds the dawn, sleep on. For Demeter has imposed the labours of your hands on the nymphs, who leaping down upon the topmost part of the wheel, rotate its axle; with encircling cogs, it turns the hollow weight of the Nisyrian millstones. If we learn to feast toil-free on the fruits of the earth, we taste again the golden age.

The Roman encyclopedist Pliny mentions in his *Naturalis Historia* of around 70 AD water-powered trip hammers operating in the greater part of Italy. There is evidence of a fulling mill in 73/4 AD in Antioch, Roman Syria.

It is likely that a water-powered stamp mill was used at Dolaucothi to crush gold-bearing quartz, with a possible date of the late 1st century to the early 2nd century. The stamps were operated as a batch of four working against a large conglomerate block, now known as *Carreg Pumpsaint*. Similar anvil stones have been found at other Roman mines across Europe, especially in Spain and Portugal.

The 1st-century AD multiple mill complex of Barbegal in southern France has been described as "the greatest known concentration of mechanical power in the ancient world". It featured 16 overshot waterwheels to power an equal number of flour mills. The capacity of the mills has been estimated at 4.5 tons of flour per day, sufficient to supply enough bread for the 12,500 inhabitants occupying the town of Arelate at that time. A similar mill complex existed on the Janiculum hill, whose supply of flour for Rome's population was judged by emperor Aurelian important enough to be included in the Aurelian walls in the late 3rd century.

A breastshot wheel mill dating to the late 2nd century AD was excavated at Les Martres-de-Veyre, France.

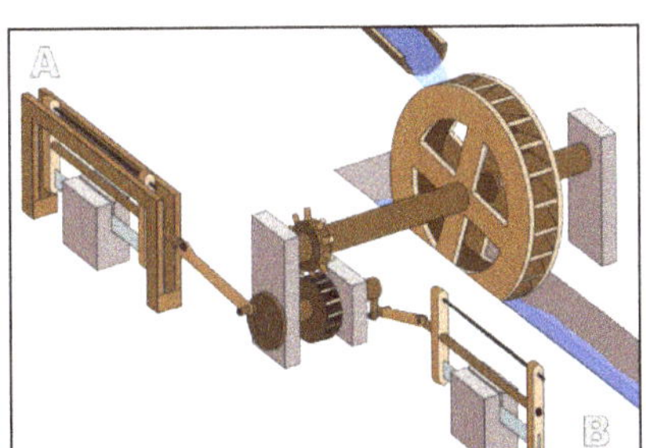

Scheme of the Roman Hierapolis sawmill, the earliest known machine to incorporate a crank and connecting rod mechanism.

The 3rd-century AD Hierapolis water-powered stone sawmill is the earliest known machine to incorporate a crank and connecting rod mechanism. Further sawmills, also powered by crank and connecting rod mechanisms, are archaeologically attested for the 6th-century water-powered stone sawmills at Gerasa and Ephesus. Literary references to water-powered marble saws in what is now Germany can be found in Ausonius 4th-century poem *Mosella*. They also seem to be indicated about the same time by the Christian saint Gregory of Nyssa from Anatolia, demonstrating a diversified use of water-power in many parts of the Roman Empire.

Roman turbine mill at Chemtou, Tunisia. The tangential water inflow of the millrace made the horizontal wheel in the shaft turn like a true turbine, the earliest known.

The earliest turbine mill was found in Chemtou and Testour, Roman North Africa, dating to the late 3rd or early 4th century AD. A possible water-powered furnace has been identified at Marseille, France.

Mills were commonly used for grinding grain into flour (attested by Pliny the Elder), but industrial uses as fulling and sawing marble were also applied.

The Romans used both fixed and floating water wheels and introduced water power to other provinces of the Roman Empire. So-called 'Greek Mills' used water wheels with a horizontal wheel (and vertical shaft). A "Roman Mill" features a vertical wheel (on a horizontal shaft). Greek style mills are the older and simpler of the two designs, but only operate well with high water velocities and with small diameter millstones. Roman style mills are more complicated as they require gears to transmit the power from a shaft with a horizontal axis to one with a vertical axis.

Although to date only a few dozen Roman mills are archaeologically traced, the widespread use of aqueducts in the period suggests that many remain to be discovered. Recent excavations in Roman London, for example, have uncovered what appears to be a tide mill together with a possible sequence of mills worked by an aqueduct running along the side of the River Fleet.

In 537 AD, ship mills were ingeniously used by the East Roman general Belisarius, when the besieging Goths cut off the water supply for those mills. These floating mills had a wheel that was attached to a boat moored in a fast flowing river.

Middle Ages

At the time of the compilation of the Domesday Book (1086), there were 5,624 watermills in England alone, only 2% of which have not been located by modern archeological surveys. Later research estimates a less conservative number of 6,082, and it has been pointed out that this should be considered a minimum as the northern reaches of England were never properly recorded. In 1300, this number had risen to between 10,000 and 15,000. By the early 7th century, watermills were well established in Ireland, and began to spread from the former territory of the empire into the non-romanized parts of Germany a century later. Ship mills and tide mill were introduced in the 6th century.

Tide mills

In recent years, a number of new archaeological finds has consecutively pushed back the date of the earliest tide mills, all of which were discovered on the Irish coast: A 6th-century vertical-wheeled tide mill was located at Killoteran near Waterford. A twin flume horizontal-wheeled tide mill dating to c. 630 was excavated on Little Island. Alongside it, another tide mill was found which was powered by a vertical undershot wheel. The Nendrum Monastery mill from 787 was situated on an island in Strangford Lough in Northern Ireland. Its millstones are 830mm in diameter and the horizontal wheel is estimated to have developed 7–8 hp at its peak. Remains of an earlier mill dated at 619 were also found at the site.

Survey of industrial mills

In a 2005 survey the scholar Adam Lucas identified the following first appearances of various industrial mill types in Western Europe. Noticeable is the preeminent role of France in the introduction of new innovative uses of waterpower. However, he has drawn attention to the dearth of studies of the subject in several other countries.

First Appearance of Various Industrial Mills in Medieval Europe, AD 770-1443		
Type of mill	Date	Country
Malt mill	770	France
Fulling mill	1080	France
Tanning mill	c. 1134	France
Forge mill	c. 1200	England, France
Tool-sharpening mill 1203 France		
Hemp mill	1209	France
Paper mill	1238, 1273	Xativa, Spain
Bellows	1269, 1283	Medieval Hungary, France
Sawmill	c. 1300	France
Ore-crushing mill	1317	Germany
Blast furnace	1384	France
Cutting and slitting mill	1443	France

Ancient China

A Northern Song era (960–1127) water-powered mill for dehusking grain with a horizontal wheel.

The waterwheel was found in China from 30 AD onwards, when it was used to power trip hammers, the bellows in smelting iron, and in one case, to mechanically rotate an armillary sphere for astronomical observation. Although Joseph Needham speculates that the water-powered millstone could have existed in Han China by the 1st century AD, there is no sufficient literary evidence for it until the 5th century. In 488 AD, the mathematician and engineer Zu Chongzhi had a watermill erected which was inspected by Emperor Wu of Southern Qi (r. 482–493 AD). The engineer Yang Su of the Sui Dynasty (581–618 AD) was said to operate hundreds of them by the beginning of the 6th century. A source written in 612 AD mentions Buddhist monks arguing over the revenues gained from watermills. The Tang Dynasty (618–907 AD) 'Ordinances of the Department of Waterways' written in 737 AD stated that watermills should not interrupt riverine transport and in some cases were restricted to use in certain seasons of the year. From other Tang-era sources of the 8th century, it is known that these ordinances were taken very seriously, as the government demolished many watermills owned by great families, merchants, and Buddhist abbeys that failed to acknowledge ordinances or meet government regulations. A eunuch serving Emperor Xuanzong of Tang (r. 712–756 AD) owned a watermill by 748 AD which employed five waterwheels that ground 300 bushels of wheat a day. By 610 or 670 AD, the watermill was introduced to Japan via Korean Peninsula. It also became known in Tibet by at least 641 AD.

Ancient India

According to Greek historical tradition, India received water-mills from the Roman Empire in the early 4th century AD when a certain Metrodoros introduced "water-mills and baths, unknown among them [the Brahmans] till then".

Islamic World

Muslim engineers adopted the Greek watermill technology from the Byzantine Empire, where it had been applied for centuries in those provinces conquered by the Muslims, including modern-day Syria, Jordan, Israel, Algeria, Tunisia, Morocco, and Spain.

The industrial uses of watermills in the Islamic world date back to the 7th century, while horizontal-wheeled and vertical-wheeled watermills were both in widespread use by the 9th century. A variety of industrial watermills were used in the Islamic world, including gristmills, hullers, sawmills, shipmills, stamp mills, steel mills, sugar mills, and tide mills. By the 11th century, every province throughout the Islamic world had these industrial watermills in operation, from al-Andalus and North Africa to the Middle East and Central Asia. Muslim and Middle Eastern Christian engineers also used crankshafts and water turbines, gears in watermills and water-raising machines, and dams as a source of water, used to provide additional power to watermills and water-raising machines. Fulling mills, and steel mills may have spread from Al-Andalus to Christian Spain in the 12th century. Industrial watermills were also employed in large factory complexes built in al-Andalus between the 11th and 13th centuries.

An Afghan water mill photographed during the Second Anglo-Afghan War (1878-1880). The rectangular water mill has a thatched roof and traditional design with a small horizontal mill-house built of stone or perhaps mud bricks.

The engineers of the Islamic world used several solutions to achieve the maximum output from a watermill. One solution was to mount them to piers of bridges to take advantage of the increased flow. Another solution was the shipmill, a type of watermill powered by water wheels mounted on the sides of ships moored in midstream. This technique was employed along the Tigris and Euphrates rivers in 10th-century Iraq, where large shipmills made of teak and iron could produce 10 tons of flour from corn every day for the granary in Baghdad.

Persia

More than 300 watermills were at work in Iran till 1960. Now only a few are still working. One of the famous ones is the water mill of Askzar and the water mill of the Yazd city, still producing flour.

Operation

A watermill in Tapolca, Veszprem County, Hungary.

Roblin's Mill, a watermill, at Black Creek Pioneer Village in Toronto, Canada.

Watermills in Bosnia and Herzegovina.

Typically, water is diverted from a river or impoundment or mill pond to a turbine or water wheel, along a channel or pipe (variously known as a flume, head race, mill race, leat, leet, lade ((Scots) or penstock). The force of the water's movement drives the blades of a wheel or turbine, which in turn rotates an axle that drives the mill's other machinery. Water leaving the wheel or turbine is drained through a tail race, but this channel may also be the head race of yet another wheel, turbine or mill. The passage of water is controlled by sluice gates that allow maintenance and some measure of flood control; large mill complexes may have dozens of sluices controlling complicated interconnected races that feed multiple buildings and industrial processes.

Watermills can be divided into two kinds, one with a horizontal water wheel on a vertical axle, and the other with a vertical wheel on a horizontal axle. The oldest of these were horizontal mills in which the force of the water, striking a simple paddle wheel set horizontally in line with the flow turned a runner stone balanced on the rynd which is atop a shaft leading directly up from the wheel. The bedstone does not turn. The problem with this type of mill arose from the lack of gearing; the speed of the water directly set the maximum speed of the runner stone which, in turn, set the rate of milling.

Most watermills in Britain and the United States of America had a vertical waterwheel, one of four kinds: undershot, breast-shot, overshot and pitchback wheels. This vertical produced rotary motion around a horizontal axis, which could be used (with cams) to lift hammers in a forge, fulling stocks in a fulling mill and so on.

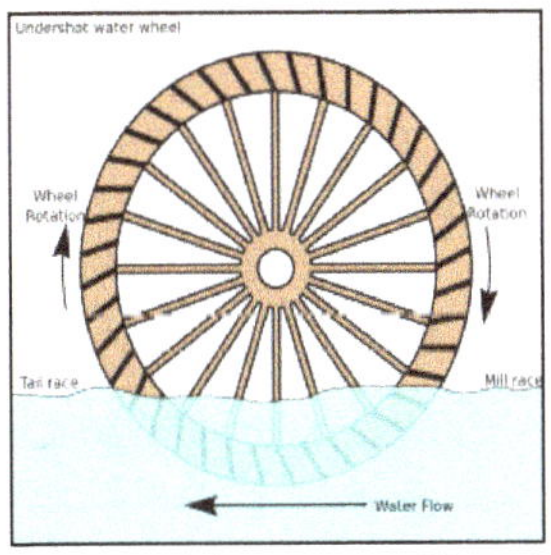

Undershot water wheel, applied for watermilling since the 1st century BC.

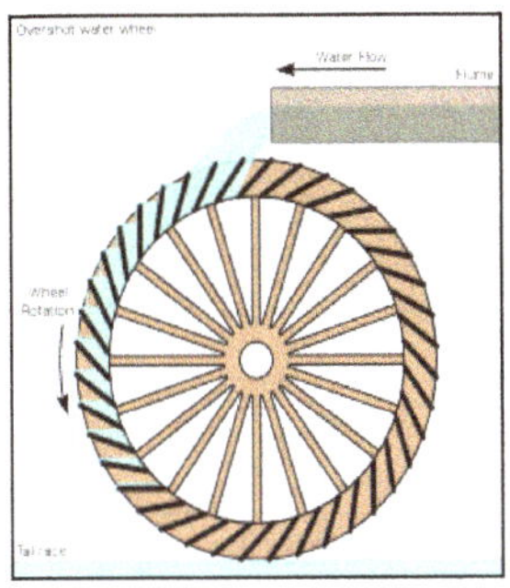

Overshot water wheel, applied for watermilling since the 1st century BC.

Milling Corn

However, in corn mills rotation about a vertical axis was required to drive its stones. The horizontal rotation was converted into the vertical rotation by means of gearing, which also enabled the runner stones to turn faster than the waterwheel. The usual arrangement in British and American corn mills has been for the waterwheel to turn a horizontal shaft on which is also mounted a large pit wheel. This meshes with the wallower, mounted on a vertical shaft, which turns the (larger) great spur wheel (mounted on the same shaft). This large face wheel, set with pegs, in turn, turned a smaller wheel (such as a lantern gear) known as a stone nut, which was attached to the shaft that drove the runner stone. The number of runner stones that could be turned depended directly upon the supply of water available. As waterwheel technology improved mills became more efficient, and by the 19th century, it was common for the great spur wheel to drive several stone nuts, so that a single water wheel could drive as many as four stones. Each step in the process increased the gear ratio which increased the maximum speed of the runner stone. Adjusting the sluice gate and thus the flow of the water past the main wheel allowed the miller to compensate for seasonal variations in the water supply. Finer speed adjustment was made during the milling process by *tentering*, that is, adjusting the gap between the stones according to the water flow, the type of grain being milled, and the grade of flour required.

In many mills (including the earliest) the great spur wheel turned only one stone, but there might be several mills under one roof. The earliest illustration of a single waterwheel driving more than one set of stones was drawn by Henry Beighton in 1723 and published in 1744 by J. T. Desaguliers.

Dalgarven Mill, Ayrshire, United Kingdom.

Shipmill on the Mura, Slovenia.

Overshot and Pitchback Mills

The overshot wheel was a later innovation in waterwheels and was around two and a half times more efficient than the undershot. The undershot wheel, in which the main water wheel is simply set into the flow of the mill race, suffers from an inherent inefficiency stemming from the fact that the wheel itself, entering the water behind the main thrust of the flow driving the wheel, followed by the lift of the wheel out of the water ahead of the main thrust, actually impedes its own operation. The overshot wheel solves this problem by bringing the water flow to the top of the wheel. The water fills buckets built into the wheel, rather than the simple paddle wheel design of undershot wheels. As the buckets fill, the weight of the water starts to turn the wheel. The water spills out of the bucket on the down side into a spillway leading back to river. Since the wheel itself is set above the spillway, the water never impedes the speed of the wheel. The impulse of the water on the wheel is also harnessed in addition to the weight of the water once in the buckets. Overshot wheels require the construction of a dam on the river above the mill and a more elaborate millpond, sluice gate, mill race and spillway or tailrace.

An inherent problem in the overshot mill is that it reverses the rotation of the wheel. If a miller wishes to convert a breastshot mill to an overshot wheel all the machinery in the mill has to be rebuilt to take account of the change in rotation. An alternative solution was the pitchback or backshot wheel. A launder was placed at the end of the flume on the headrace, this turned the direction of the water without much loss of energy, and the direction of rotation was maintained. Daniels Mill near Bewdley, Worcestershire is an example of a flour mill that originally used a breastshot wheel, but was converted to use a pitchback wheel. Today it operates as a breastshot mill.

A Breastshot waterwheel at Dalgarven Mill, United Kingdom.

Larger water wheels (usually overshot steel wheels) transmit the power from a toothed annular ring that is mounted near the outer edge of the wheel. This drives the machinery using a spur gear mounted on a shaft rather than taking power from the central axle. However, the basic mode of operation remains the same; gravity drives machinery through the motion of flowing water.

Toward the end of the 19th century, the invention of the Pelton wheel encouraged some mill owners to replace over- and undershot wheels with Pelton wheel turbines driven through penstocks.

Tide Mills

A different type of watermill is the tide mill. This mill might be of any kind, undershot, overshot or horizontal but it does not employ a river for its power source. Instead a mole or causeway is built across the mouth of a small bay. At low tide, gates in the mole are opened allowing the bay to fill with the incoming tide. At high tide the gates are closed, trapping the water inside. At a certain point a sluice gate in the mole can be opened allowing the draining water to drive a mill wheel or wheels. This is particularly effective in places where the tidal differential is very great, such as the Bay of Fundy in Canada where the tides can rise fifty feet, or the now derelict village of Tide Mills in the United Kingdom. A working example can be seen at Eling Tide Mill.

Run of the river schemes do not divert water at all and usually involve undershot wheels the mills are mostly on the banks of sizeable rivers or fast flowing streams. Other watermills were set beneath large bridges where the flow of water between the stanchions was faster. At one point London bridge had so many water wheels beneath it that bargemen complained that passage through the bridge was impaired.

Current Status

By the early 20th century, availability of cheap electrical energy made the watermill obsolete in developed countries although some smaller rural mills continued to operate commercially later throughout the century. A few historic mills such as the Newlin Mill and Yates Mill in the USA and The Darley Mill Centre in the UK still operate for demonstration purposes. Small-scale commercial production is carried out in the UK at Daniels Mill, Little Salkeld Mill and Redbournbury Mill.

Watermill in Jahodná (Slovakia).

Some old mills are being upgraded with modern Hydropower technology, for example those worked on by the South Somerset Hydropower Group in the UK.

In some developing countries, watermills are still widely used for processing grain. For example, there are thought to be 25,000 operating in Nepal, and 200,000 in India. Many of these are still of the traditional style, but some have been upgraded by replacing wooden parts with better-designed metal ones to improve the efficiency. For example, the Centre for Rural Technology in Nepal upgraded 2,400 mills between 2003 and 2007.

Applications

Watermill in Caldas Novas, Brazil.

Former watermill in Kohila, Estonia.

- Bark mills ground bark, from oak or chestnut trees to produce a coarse powder for use in tanneries.
- Blade mills were used for sharpening newly made blades.
- Blast furnaces, finery forges, and tinplate works were, until the introduction of the steam

engine, almost invariably water powered. Furnaces and Forges were sometimes called iron mills.

- Bobbin mills made wooden bobbins for the cotton and other textile industries.
- Carpet mills for making carpets and rugs were sometimes water-powered.
- Cotton mills were driven by water. The power was used to card the raw cotton, and then to drive the spinning mules and ring frames. Steam engines were initially used to increase the water flow to the wheel, then as the industrial revolution progressed, to directly drive the shafts.
- Fulling or *walk* mills were used for a finishing process on woollen cloth.
- Gristmills, or *corn mills*, grind grains into flour.
- Lead was usually smelted in smeltmills prior to the introduction of the cupola (a reverberatory furnace).
- Needle mills for scouring needles during manufacture were mostly water-powered (such as Forge Mill Needle Museum).
- Oil mills for crushing oil seeds might be wind or water-powered.
- Paper mills used water not only for motive power, but also required it in large quantities in the manufacturing process.
- Powder mills for making gunpowder - black powder or smokeless powder were usually water-powered.
- Rolling mills shaped metal by passing it between rollers.
- Sawmills cut timber into lumber.
- Slitting mills were used for slitting bars of iron into rods, which were then made into nails.
- Spoke mills turned lumber into spokes for carriage wheels.
- Stamp mills for crushing ore, usually from non-ferrous mines.
- Textile mills for spinning yarn or weaving cloth were sometimes water-powered.

Water Wheel

A water wheel is a machine for converting the energy of free-flowing or falling water into useful forms of power, often in a watermill. A water wheel consists of a large wooden or metal wheel, with a number of blades or buckets arranged on the outside rim forming the driving surface. Most commonly, the wheel is mounted vertically on a horizontal axle, but the tub or Norse wheel is mounted horizontally on a vertical shaft. Vertical wheels can transmit power either through the axle or via a ring gear and typically drive belts or gears; horizontal wheels usually directly drive their load.

Water wheels were still in commercial use well into the 20th century, but they are no longer in common use. Prior uses of water wheels include milling flour in gristmills and grinding wood into pulp for papermaking, but other uses include hammering wrought iron, machining, ore crushing and pounding fiber for use in the manufacture of cloth.

An overshot waterwheel standing 42 ft (13 m) high powers the Old Mill at Berry College in Rome, Georgia, USA.

Water wheel powering a mine hoist in *De re metallica* (1566).

Some water wheels are fed by water from a mill pond, which is formed when a flowing stream is dammed. A channel for the water flowing to or from a water wheel is called a mill race (also spelled millrace) or simply a "race", and is customarily divided into sections. The race bringing water from the mill pond to the water wheel is a headrace; the one carrying water after it has left the wheel is commonly referred to as a tailrace.

John Smeaton's scientific investigation of the water wheel led to significant increases in efficiency in the mid to late 18th century and supplying much needed power for the Industrial Revolution.

Water wheels began being displaced by the smaller, less expensive and more efficient turbine, developed by Benoît Fourneyron, beginning with his first model in 1827. Turbines are capable of handling high *heads*, or elevations, that exceed the capability of practical-sized waterwheels.

The main difficulty of water wheels is their dependence on flowing water, which limits where they can be located. Modern hydroelectric dams can be viewed as the descendants of the water wheel, as they too take advantage of the movement of water downhill.

History

The two main functions of water wheels were historically water-lifting for irrigation purposes and as a power source. In terms of power source, water wheels can be turned either by human or animal force or by the water current itself. Water wheels come in two basic designs, either equipped with a vertical or a horizontal axle. The latter type can be subdivided, depending on where the water hits the wheel paddles, into overshot, breastshot and undershot wheels.

Greco-Roman World

Engineers of the Hellenistic era Mediterranean region are credited with the development of the water wheel. Mediterranean engineers of the Hellenistic and Roman periods were also the first to use it for both irrigation and as a power source. The technological breakthrough occurred in the technically advanced and scientifically minded Hellenistic period between the 3rd and 1st centuries BCE. This is seen as an evolution of the paddle-driven water-lifting wheels that had appeared in ancient Egypt by the 4th century BCE. According to John Peter Oleson, both the compartmented wheel and the hydraulic Noria appeared in Egypt by the 4th century BCE, with the Sakia being invented there a century later. This is supported by archeological finds at Faiyum, where the oldest archeological evidence of a water-wheel has been found, in the form of a Sakia dating back to the 3rd century BCE. A papyrus dating to the 2nd century BCE also found in Faiyum mentions a water wheel used for irrigation, a 2nd-century BC fresco found at Alexandria depicts a compartmented Sakia, and the writings of Callixenus of Rhodes mention the use of a Sakia in Ptolemaic Egypt during the reign of Ptolemy IV in the late 3rd century BC.

Drainage Wheels

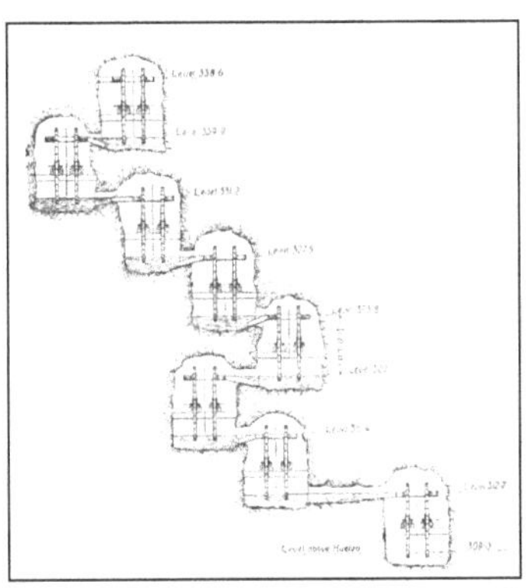

Sequence of wheels found in Rio Tinto mines, southwestern Spain.

The Romans used water wheels extensively in mining projects. Several such devices were described by Vitruvius. The one found during modern mining at the copper mines at Rio Tinto in Spain involved 16 such wheels stacked above one another so as to lift water about 80 feet (24 m) from the mine sump. Part of a similar wheel dated to about 90 CE, was found in the 1930s, at Dolaucothi, a Roman gold mine in south Wales.

Water Mills

Reconstruction of Vitruvius' undershot-wheeled watermill.

Taking indirect evidence into account from the work of the Greek technician Apollonius of Perge, the British historian of technology M.J.T. Lewis dates the appearance of the vertical-axle watermill to the early 3rd century BCE, and the horizontal-axle watermill to around 240 BC, with Byzantium and Alexandria as the assigned places of invention. A watermill is reported by the Greek geographer Strabon (ca. 64 BCE–CE 24) to have existed sometime before 71 BCE in the palace of the Pontian king Mithradates VI Eupator, but its exact construction cannot be gleaned from the text (XII, 3, 30 C 556).

The first clear description of a geared watermill is from the 1st-century BC Roman architect Vitruvius, who tells of the sakia gearing system as being applied to a watermill. Vitruvius's account is particularly valuable in that it shows how the watermill came about, namely by the combination of the separate Greek inventions of the toothed gear and the water wheel into one effective mechanical system for harnessing water power. Vitruvius's water wheel is described as being immersed with its lower end in the watercourse so that its paddles could be driven by the velocity of the running water (X, 5.2).

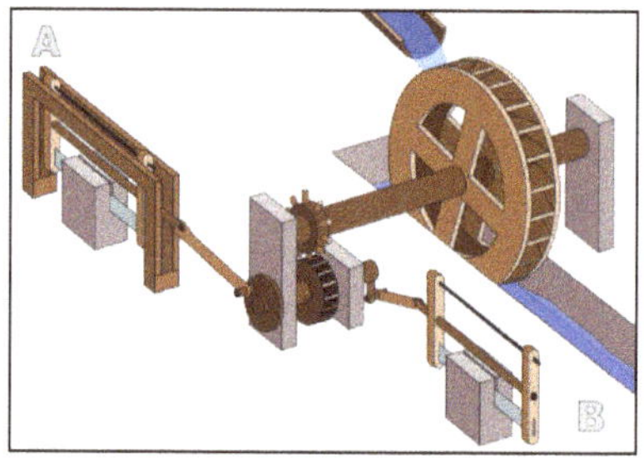

Schematic of the Roman Hierapolis sawmill, Asia Minor, powered by a breastshot wheel.

About the same time, the overshot wheel appears for the first time in a poem by Antipater of Thessalonica, which praises it as a labour-saving device (IX, 418.4–6). The motif is also taken up by

Lucretius (ca. 99-55 BC) who likens the rotation of the water wheel to the motion of the stars on the firmament (V 516). The third horizontal-axled type, the breastshot water wheel, comes into archaeological evidence by the late-2nd-century AD context in central Gaul. Most excavated Roman watermills were equipped with one of these wheels which, although more complex to construct, were much more efficient than the vertical-axle water wheel. In the 2nd century AD, Barbegal watermill complex a series of sixteen overshot wheels was fed by an artificial aqueduct, a proto-industrial grain factory which has been referred to as "the greatest known concentration of mechanical power in the ancient world".

In Roman North Africa, several installations from around 300 AD were found where vertical-axle water wheels fitted with angled blades were installed at the bottom of a water-filled, circular shaft. The water from the mill-race which entered the pit tangentially created a swirling water column that made the fully submerged wheel act like true water turbines, the earliest known to date.

Navigation

Ox-powered Roman paddle wheel boat from a 15th-century copy of *De Rebus Bellicis*.

Apart from its use in milling and water-raising, ancient engineers applied the paddled water wheel for automatons and in navigation. Vitruvius (X 9.5-7) describes multi-geared paddle wheels working as a ship odometer, the earliest of its kind. The first mention of paddle wheels as a means of propulsion comes from the 4th–5th-century military treatise *De Rebus Bellicis* (chapter XVII), where the anonymous Roman author describes an ox-driven paddle-wheel warship.

Early Medieval Europe

Ancient water-wheel technology continued unabated in the early medieval period where the appearance of new documentary genres such as legal codes, monastic charters, but also hagiography was accompanied with a sharp increase in references to watermills and wheels.

The earliest vertical-wheel in a tide mill is from 6th-century Killoteran near Waterford, Ireland, while the first known horizontal-wheel in such a type of mill is from the Irish Little Island (c. 630). As for the use in a common Norse or Greek mill, the oldest known horizontal-wheels were excavated in the Irish Ballykilleen, dating to c. 636.

The earliest excavated water wheel driven by tidal power was the Nendrum Monastery mill in Northern Ireland which has been dated at 787A.D. although a possible earlier mill dates to 619A.D. Tide mills became common in estuaries with a good tidal range in both Europe and America generally using undershot wheels.

Water wheel powering a small village mill at the Museum of Folk Architecture and Life, Uzhhorod, Ukraine.

Cistercian monasteries, in particular, made extensive use of water wheels to power watermills of many kinds. An early example of a very large water wheel is the still extant wheel at the early 13th century Real Monasterio de Nuestra Senora de Rueda, a Cistercian monastery in the Aragon region of Spain. Grist mills (for corn) were undoubtedly the most common, but there were also sawmills, fulling mills and mills to fulfil many other labour-intensive tasks. The water wheel remained competitive with the steam engine well into the Industrial Revolution. At around the 8th to 10th century, a number of irrigation technologies were brought into Spain and thus introduced to Europe. One of those technologies is the Noria, which is basically a wheel fitted with buckets on the peripherals for lifting water. It is similar to the undershot water wheel mentioned later in this article. It allowed peasants to power watermills more efficiently. According to Thomas Glick's book, *Irrigation and Society in Medieval Valencia*, the Noria probably originated from somewhere in Persia. It has been used for centuries before the technology was brought into Spain by Arabs who had adopted it from the Romans. Thus the distribution of the Noria in the Iberian peninsula "conforms to the area of stabilized Islamic settlement". This technology has a profound effect on the life of peasants. The Noria is relatively cheap to build. Thus it allowed peasants to cultivate land more efficiently in Europe. Together with the Spaniards, the technology then spread to North Africa and later to the New World in Mexico and South America following Spanish expansion.

Domesday Inventory of English Mills Ca. 1086

The assembly convened by William of Normandy, commonly referred to as the "Domesday" or Doomsday survey, took an inventory of all potentially taxable property in England, which included over six thousand mills spread across three thousand different locations.

Locations

The type of water wheel selected was dependent upon the location. Generally if only small volumes of water and high waterfalls were available a millwright would choose to use an overshot wheel. The decision was influenced by the fact that the buckets could catch and use even a small volume of water. For large volumes of water with small waterfalls the undershot wheel would have been used, since it was more adapted to such conditions and cheaper to construct. So long as these water supplies were abundant the question of efficiency remained irrelevant. By the 18th century with increased demand for power coupled with limited water locales, an emphasis was made on efficiency scheme.

Economic Influence

By the 11th century there were parts of Europe where the exploitation of water was commonplace. The water wheel is understood to have actively shaped and forever changed the outlook of Westerners. Europe began to transit from human and animal muscle labor towards mechanical labor with the advent of the water wheel. Medievalist Lynn White Jr. contended that the spread of inanimate power sources was eloquent testimony to the emergence of the West of a new attitude toward, power, work, nature, and above all else technology.

Harnessing water-power enabled gains in agricultural productivity, food surpluses and the large scale urbanization starting in the 11th century. The usefulness of water power motivated European experiments with other power sources, such as wind and tidal mills. Waterwheels influenced the construction of cities, more specifically canals. The techniques that developed during this early period such as stream jamming and the building of canals, put Europe on a hydraulically focused path, for instance water supply and irrigation technology was combined to modify supply power of the wheel. Illustrating the extent to which there was a great degree of technological innovation that met the growing needs of the feudal state.

Applications of the Water Wheel in Medieval Europe

Ore stamp mill (behind worker taking ore form chute).

The water mill was used for grinding grain, producing flour for bread, malt for beer, or coarse meal for porridge. Hammermills used the wheel to operate hammers. One type was fulling mill, which was used for cloth making. The trip hammer was also used for making wrought iron and for working iron into useful shapes, an activity that was otherwise labour-intensive. The water wheel was also used in papermaking, beating material to a pulp. In the 13th century water mills used for hammering throughout Europe improved the productivity of early steel manufacturing. Along with the mastery of gunpowder, waterpower provided European countries worldwide military leadership from the 15th century.

Importance to 17th- and 18th-century Europe (Scientific Influence)

Millwrights distinguished between the two forces, impulse and weight, at work in water wheels long before 18th-century Europe. Fitzherbert, a 16th-century agricultural writer, wrote "druieth the wheel as well as with the weight of the water as with strengthe [impulse].” Leonardo da Vinci also discussed water power, noting “the blow [of the water] is not weight, but excites a power of weight, almost equal to its own power.” However, even realisation of the two forces, weight and impulse, confusion remained over the advantages and disadvantages of the two, and there was no clear understanding of the superior efficiency of weight. Prior to 1750 it was unsure as to which force was dominant and was widely understood that both forces were operating with equal inspiration amongst one another. The waterwheel, sparked questions of the laws of nature, specifically the laws of force. Evangelista Torricelli’s work on water wheels used an analysis of Galileo’s work on falling bodies, that the velocity of a water sprouting from an orifice under its head was exactly equivalent to the velocity a drop of water acquired in falling freely from the same height.

Industrial European Usage

Lady Isabella Wheel, Laxey, Isle of Man, used to drive mine pumps.

A mid-19th-century water wheel at Cromford in England used for grinding locally mined barytes.

The most powerful water wheel built in the United Kingdom was the 100 hp Quarry Bank Mill water wheel near Manchester. A high breastshot design, it was retired in 1904 and replaced with several turbines. It has now been restored and is a museum open to the public.

The biggest working water wheel in mainland Britain has a diameter of 15.4 m and was built by the De Winton company of Caernarfon. It is located within the Dinorwic workshops of the National Slate Museum in Llanberis, North Wales.

The largest working water wheel in the world is the Laxey Wheel (also known as *Lady Isabella*) in the village of Laxey, Isle of Man. It is 72 feet 6 inches (22.10 m) in diameter and 6 feet (1.83 m) wide and is maintained by Manx National Heritage.

Development of water turbines during the Industrial Revolution led to decreased popularity of water wheels. The main advantage of turbines is that its ability to harness head is much greater than the diameter of the turbine, whereas a water wheel cannot effectively harness head greater than its diameter. The migration from water wheels to modern turbines took about one hundred years.

China

Two types of hydraulic-powered chain pumps.

Chinese water wheels almost certainly have a separate origin, as early ones there were invariably horizontal water wheels. By at least the 1st century AD, the Chinese of the Eastern Han Dynasty were using water wheels to crush grain in mills and to power the piston-bellows in forging iron ore into cast iron.

In the text known as the *Xin Lun* written by Huan Tan about 20 AD (during the usurpation of Wang Mang), it states that the legendary mythological king known as Fu Xi was the one responsible for the pestle and mortar, which evolved into the tilt-hammer and then trip hammer device. Although the author speaks of the mythological Fu Xi, a passage of his writing gives hint that the water wheel was in widespread use by the 1st century AD in China (Wade-Giles spelling):

Fu Hsi invented the pestle and mortar, which is so useful, and later on it was cleverly improved in such a way that the whole weight of the body could be used for treading on the tilt-hammer (*tui*), thus increasing the efficiency ten times. Afterwards the power of animals—donkeys, mules, oxen, and horses—was applied by means of machinery, and water-power too used for pounding, so that the benefit was increased a hundredfold.

In the year 31 AD, the engineer and Prefect of Nanyang, Du Shi (d. 38), applied a complex use of the water wheel and machinery to power the bellows of the blast furnace to create cast iron. Du Shi is mentioned briefly in the *Book of Later Han* (*Hou Han Shu*) as follows (in Wade-Giles spelling):

In the seventh year of the Chien-Wu reign period (31 AD) Tu Shih was posted to be Prefect of Nanyang. He was a generous man and his policies were peaceful; he destroyed evil-doers and established the dignity (of his office). Good at planning, he loved the common people and wished to save their labor. He invented a water-power reciprocator (*shui phai*) for the casting of (iron) agricultural implements. Those who smelted and cast already had the push-bellows to blow up their charcoal fires, and now they were instructed to use the rushing of the water (*chi shui*) to operate it ... Thus the people got great benefit for little labor. They found the 'water(-powered) bellows' convenient and adopted it widely.

Water wheels in China found practical uses such as this, as well as extraordinary use. The Chinese inventor Zhang Heng (78–139) was the first in history to apply motive power in rotating the astronomical instrument of an armillary sphere, by use of a water wheel. The mechanical engineer Ma Jun (c. 200–265) from Cao Wei once used a water wheel to power and operate a large mechanical puppet theater for the Emperor Ming of Wei (r. 226-239).

India

The early history of the watermill in India is obscure. Ancient Indian texts dating back to the 4th century BC refer to the term *cakkavattaka* (turning wheel), which commentaries explain as *arahatta-ghati-yanta* (machine with wheel-pots attached). On this basis, Joseph Needham suggested that the machine was a noria. Terry S. Reynolds, however, argues that the "term used in Indian texts is ambiguous and does not clearly indicate a water-powered device." Thorkild Schiøler argued that it is "more likely that these passages refer to some type of tread- or hand-operated water-lifting device, instead of a water-powered water-lifting wheel."

According to Greek historical tradition, India received water-mills from the Roman Empire in the early 4th century AD when a certain Metrodoros introduced "water-mills and baths, unknown among them [the Brahmans] till then". Irrigation water for crops was provided by using water raising wheels, some driven by the force of the current in the river from which the water was being raised. This kind of water raising device was used in ancient India, predating, according to Pacey, its use in the later Roman Empire or China, even though the first literary, archaeological and pictorial evidence of the water wheel appeared in the Hellenistic world.

Around 1150, the astronomer Bhaskara Achārya observed water-raising wheels and imagined such a wheel lifting enough water to replenish the stream driving it, effectively, a perpetual motion machine. The construction of water works and aspects of water technology in India is described in Arabic and Persian works. During medieval times, the diffusion of Indian and Persian irrigation technologies gave rise to an advanced irrigation system which brought about economic growth and also helped in the growth of material culture.

Islamic World

Arab engineers took over the water technology of the hydraulic societies of the ancient Near East; they adopted the Greek water wheel as early as the 7th century, excavation of a canal in the Basra region discovered remains of a water wheel dating from this period. Hama in Syria still preserves some of its large wheels, on the river Orontes, although they are no longer in use. One of the largest had a diameter of about 20 metres and its rim was divided into 120 compartments. Another wheel that is still in operation is found at Murcia in Spain, La Nora, and although the original wheel has been replaced by a steel one, the Moorish system during al-Andalus is otherwise virtually unchanged. Some medieval Islamic compartmented water wheels could lift water as high as 30 meters. Muhammad ibn Zakariya al-Razi's *Kitab al-Hawi* in the 10th century described a noria in Iraq that could lift as much as 153,000 litres per hour, or 2550 litres per minute. This is comparable to the output of modern norias in East Asia, which can lift up to 288,000 litres per hour, or 4800 litres per minute.

The norias of Hama on the Orontes River.

Water wheel in Djambi, Sumatra.

The industrial uses of watermills in the Islamic world date back to the 7th century, while horizontal-wheeled and vertical-wheeled water mills were both in widespread use by the 9th century. A variety of industrial watermills were used in the Islamic world, including gristmills, hullers, sawmills, shipmills, stamp mills, steel mills, sugar mills, and tide mills. By the 11th century, every province throughout the Islamic world had these industrial watermills in operation, from al-Andalus and North Africa to the Middle East and Central Asia. Muslim and Christian engineers also used crankshafts and water turbines, gears in watermills and water-raising machines, and dams as a source of water, used to provide additional power to watermills and water-raising machines. Fulling mills and steel mills may have spread from Islamic Spain to Christian Spain in the 12th century. Industrial water mills were also employed in large factory complexes built in al-Andalus between the 11th and 13th centuries.

The engineers of the Islamic world developed several solutions to achieve the maximum output from a water wheel. One solution was to mount them to piers of bridges to take advantage of the increased flow. Another solution was the shipmill, a type of water mill powered by water wheels mounted on the sides of ships moored in midstream. This technique was employed along the Tigris and Euphrates rivers in 10th-century Iraq, where large shipmills made of teak and iron could produce 10 tons of flour from corn every day for the granary in Baghdad. The flywheel mechanism, which is used to smooth out the delivery of power from a driving device to a driven machine, was invented by Ibn Bassal (fl. 1038-1075) of Al-Andalus; he pioneered the use of the flywheel in the saqiya (chain pump) and noria. The engineers Al-Jazari in the 13th century and Taqi al-Din in the 16th century described many inventive water-raising machines in their technological treatises. They also employed water wheels to power a variety of devices, including various water clocks and automata.

Types

Most water wheels in the United Kingdom and the United States are (or were) vertical wheels rotating about a horizontal axle, but in the Scottish highlands and parts of southern Europe mills often had a horizontal wheel (with a vertical axle). Water wheels are classified by the way in which water is applied to the wheel, relative to the wheel's axle. Overshot and pitchback water wheels are suitable where there is a small stream with a height difference of more than 2 meters, often in association with a small reservoir. Breastshot and undershot wheels can be used on rivers or high volume flows with large reservoirs.

Horizontal Wheel

Commonly called a tub wheel or Norse mill, the horizontal wheel is essentially a very primitive and inefficient form of the modern turbine. It is usually mounted inside a mill building below the working floor. A jet of water is directed on to the paddles of the water wheel, causing them to turn; water exits beneath the wheel, generally through the center. This is a simple system, usually used without gearing so that the vertical axle of the water wheel becomes the drive spindle of the mill.

Undershot Wheel

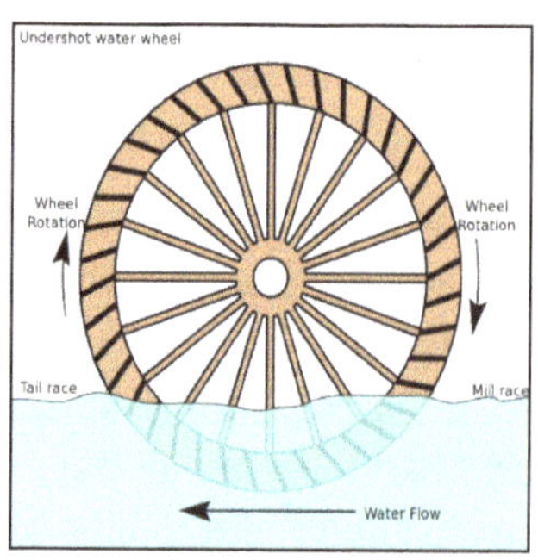

Undershot water wheel.

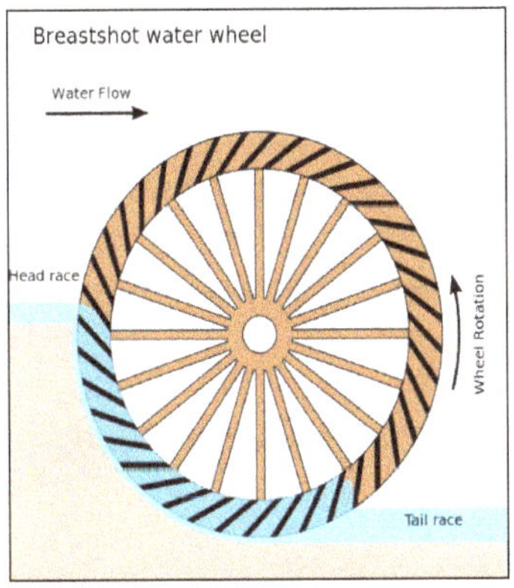

Breastshot water wheel.

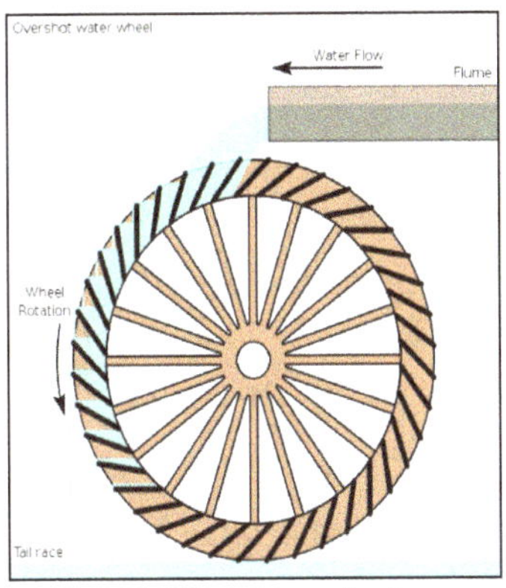

Overshot water wheel.

An undershot wheel (also called a *stream wheel*) is a vertically mounted water wheel that is rotated by water striking paddles or blades at the bottom of the wheel. The name *undershot* comes from this striking at the bottom of the wheel. This type of water wheel is the oldest type of wheel.

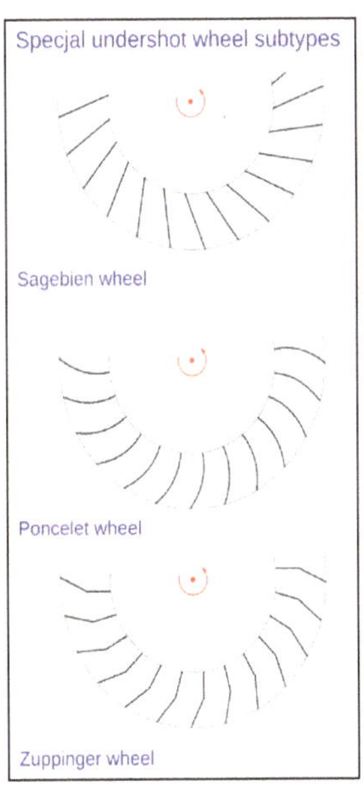

Sabegien, Poncelet and Zuppinger water wheel.

It is also regarded as the least efficient type, although subtypes of this water wheel (e.g. the Poncelet wheel, Sagebien wheel and Zuppinger wheel) allow somewhat greater efficiencies than the traditional undershot wheels. The advantages of undershot wheels are that they are somewhat cheaper and simpler to build, and have less of an environmental impact—as they do not constitute a major change of the river. Their disadvantages are—as mentioned before—less efficiency, which means that they generate less power and can only be used where the flow rate is sufficient to provide torque.

Undershot wheels gain no advantage from head. They are most suited to shallow streams in flat country.

Undershot wheels are also well suited to installation on floating platforms. The earliest were probably constructed by the Byzantine general Belisarius during the siege of Rome in 537. Later they were sometimes mounted immediately downstream from bridges where the flow restriction of arched bridge piers increased the speed of the current.

Breastshot Wheel

A vertically mounted water wheel that is rotated by falling water striking buckets near the center of the wheel's edge, or just above it, is said to be *breastshot*. Breastshot wheels are the most common type in the United States of America and are said to have powered the American industrial revolution.

Breastshot wheels are less efficient than overshot wheels, are more efficient than undershot wheels, and are not backshot. The individual blades of a breastshot wheel are actually buckets, as are those of most overshot wheels, and not simple paddles like those of most undershot wheels. A breastshot wheel requires a good trash rack and typically has a masonry "apron" closely conforming to the wheel face, which helps contain the water in the buckets as they progress downwards. Breastshot wheels are preferred for steady, high-volume flows such as are found on the fall line of the North American East Coast.

The Anderson Mill of Texas is undershot, backshot, and overshot using two sources of water. This allows the speed of the wheel to be controlled.

Overshot Wheel

A vertically mounted water wheel that is rotated by falling water striking paddles, blades or buckets near the top of the wheel is said to be *overshot*. In true overshot wheels the water passes over the top of the wheel, but the term is sometimes applied to backshot or pitchback wheels where the water goes down behind the water wheel.

A typical overshot wheel has the water channeled to the wheel at the top and slightly beyond the axle. The water collects in the buckets on that side of the wheel, making it heavier than the other "empty" side. The weight turns the wheel, and the water flows out into the tail-water when the wheel rotates enough to invert the buckets. The overshot design can use all of the water flow for power (unless there is a leak) and does not require rapid flow.

Unlike undershot wheels, overshot wheels gain a double advantage from gravity. Not only is the momentum of the flowing water partially transferred to the wheel, the weight of the water descending in the wheel's buckets also imparts additional energy. The mechanical power derived from an overshot wheel is determined by the wheel's physical size and the available head, so they are ideally suited to hilly or mountainous country. On average, the undershot wheel uses 22 percent of the energy in the flow of water, while an overshot wheel uses 63 percent, as calculated by English civil engineer John Smeaton in the 18th century.

Overshot wheels demand exact engineering and significant head, which usually means significant investment in constructing a dam, millpond and waterways. Sometimes the final approach of the water to the wheel is along a lengthy flume or penstock.

Reversible Wheel

Replica of a reversible wheel with a 9.5 m diameter in Clausthal-Zellerfeld.

A special type of overshot wheel is the reversible water wheel. This has two sets of blades or buckets running in opposite directions, so that it can turn in either direction depending on which side the water is directed. Reversible wheels were used in mining industry in order to power various means of ore conveyance. By changing the direction of the wheel, barrels or baskets of ore could be lifted up or lowered down a shaft. As a rule there was also a cable drum or a chain basket (German: *Kettenkorb*) on the axle of the wheel. It was also essential that the wheel had braking equipment in order to be able to stop the wheel (known as a braking wheel). The oldest known drawing of a reversible water wheel was by Georgius Agricola and dates to 1556.

Backshot Wheel

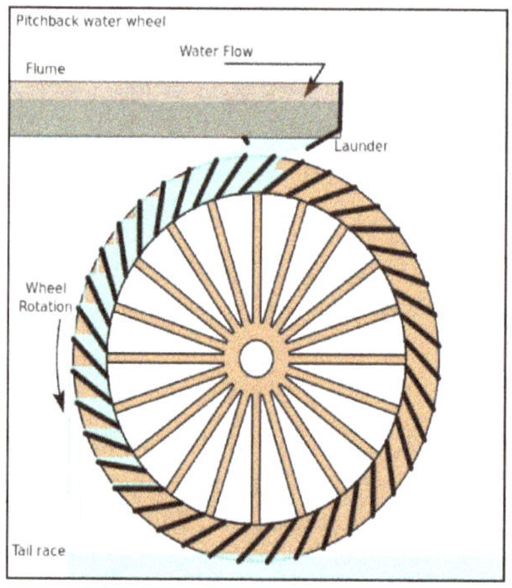

Pitchback or "backshot" water wheel.

Backshot wheel at New Lanark World Heritage Site, Scotland.

A backshot wheel (also called *pitchback*) is a variety of overshot wheel where the water is introduced just behind the summit of the wheel. It combines the advantages from breastshot and overshot systems, since the full amount of the potential energy released by the falling water is harnessed as the water descends the back of the wheel (as in overshot wheel) while it also gains power from the water's current past the bottom of the wheel (as in breastshot wheel).

A backshot wheel continues to function until the water in the wheel pit rises well above the height of the axle, when any other overshot wheel will be stopped or even destroyed. This makes the technique particularly suitable for streams that experience extreme seasonal variations in flow, and reduces the need for complex sluice and tail race configurations.

The direction of rotation of a backshot wheel is the same as that of a breastshot wheel at the same location so it can easily replace one, without causing the directional gearing in the mill to be changed. This would increase the power available while only requiring a change to be made to the water level in the top pond, which in some cases is economically viable.

Suspension Wheels and Rim-gears

Two early improvements were suspension wheels and rim gearing. Suspension wheels are constructed in the same manner as a bicycle wheel, the rim being supported under tension from the hub- this led to larger lighter wheels than the former design where the heavy spokes were under compression. Rim-gearing entailed adding a notched wheel to the rim or shroud of the wheel. A stub gear engaged the rim-gear and took the power into the mill using an independent line shaft. This removed the rotative stress from the axle which could thus be lighter, and also allowed more flexibility in the location of the power train. The shaft rotation was geared up from that of the wheel which led to less power loss. An example of this design pioneered by Thomas Hewes and refined by William Fairburn can be seen at the 1849 restored wheel at the Portland Basin Canal Warehouse.

The suspension wheel with rim-gearing at the Portland Basin Canal Warehouse.

Efficiency

Overshot (and particularly backshot) wheels are the most efficient type; a backshot steel wheel can be more efficient (about 60%) than all but the most advanced and well-constructed turbines. In some situations an overshot wheel is preferable to a turbine.

The development of the hydraulic turbine wheels with their improved efficiency (>67%) opened up an alternative path for the installation of water wheels in existing mills, or redevelopment of abandoned mills.

Power Calculations

The great water wheel in the Welsh National Slate Museum.

In an undershot wheel or a run of the river wheel the power is dependant to the kinetic energy of the river. Approximate power can be calculated.

Power in Watts= $100 \times A \times V^3 \times C$

A = Area of paddles in the water (square meters)

V = Velocity of the stream in meters per second

C = Efficiency Constant (assume 1 for a water to wire efficiency of 20%)

Rotational speed of the wheel = $9 \times V / D$ rpm

D = diameter in meters

For a breast shot or over shot wheel both potential energy and kinetic energy must be considered. This takes the form of the weight of water in the buckets and the vertical distance travelled. A rule of thumb formula is,

Power in Watts = $4 \times Q \times H \times C$

Q = Weight of water (volume per sec x capacity of the buckets)

V = Velocity of the stream in meters per second

H = Head, or height difference of water between the lip of the flume (head race) and the tailrace

C = Efficiency Constant

The optimal rotational speed of a breast shot or overshot wheel is approximately:

Rotational speed of the wheel= $21/\sqrt{D}$

D = diameter of the wheel in metres

Hydraulic Wheel

A recent development of the breastshot wheel is a hydraulic wheel which effectively incorporates automatic regulation systems. The Aqualienne is one example. It generates between 37 kW and 200 kW of electricity from a $20m^3$ waterflow with a head of 1 to 3.5m. It is designed to produce electricity at the sites of former watermills.

Hydraulic Wheel Part Reaction Turbine

A parallel development is the hydraulic wheel/part reaction turbine that also incorporates a weir into the centre of the wheel but uses blades angled to the water flow. The WICON-Stem Pressure Machine (SPM) exploits this flow. Estimated efficiency 67%.

The University of Southampton School of Civil Engineering and the Environment in the UK has investigated both types of Hydraulic wheel machines and has estimated their hydraulic efficiency and suggested improvements, i.e. The Rotary Hydraulic Pressure Machine. (Estimated maximum efficiency 85%).

These type of water wheels have high efficiency at part loads / variable flows and can operate at very low heads, < 1 metre. Combined with direct drive Axial Flux Permanent Magnet Alternators and power electronics they offer a viable alternative for low head hydroelectric power generation.

Water-lifting

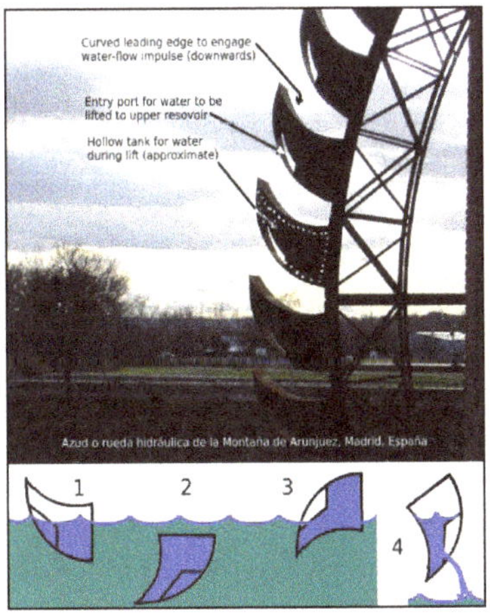

Detail of *azud* at Aranjuez, Spain.

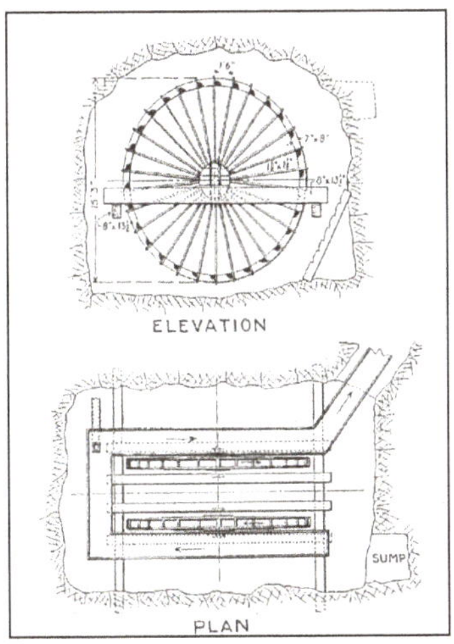

Drainage wheel from Rio Tinto mines.

In water-raising devices rotary motion is typically more efficient than machines based on oscillating motion.

The compartmented water wheel comes in two basic forms, the wheel with compartmented body (Latin *tympanum*) and the wheel with compartmented rim or a rim with separate, attached containers. The wheels could be either turned by the flow of water, men treading on its outside or by animals by means of a sakia gear. While the tympanum had a large discharge capacity, it could lift the water only to less than the height of its own radius and required a large torque for rotating. These constructional deficiencies were overcome by the wheel with a compartmented rim which was a less heavy design with a higher lift.

Ptolemaic Egypt

The earliest literary reference to a water-driven, compartmented wheel appears in the technical treatise *Pneumatica* (chap. 61) of the Greek engineer Philo of Byzantium (ca. 280–220 BC). In his *Parasceuastica* (91.43–44), Philo advises the use of such wheels for submerging siege mines as a defensive measure against enemy sapping. Compartmented wheels appear to have been the means of choice for draining dry docks in Alexandria under the reign of Ptolemy IV (221–205 BC). Several Greek papyri of the 3rd to 2nd century BC mention the use of these wheels, but don't give further details. The non-existence of the device in the Ancient Near East before Alexander's conquest can be deduced from its pronounced absence from the otherwise rich oriental iconography on irrigation practices. Unlike other water-lifting devices and pumps of the period though, the invention of the compartmented wheel cannot be traced to any particular Hellenistic engineer and may have been made in the late 4th century BC in a rural context away from the metropolis of Alexandria.

The earliest depiction of a compartmented wheel is from a tomb painting in Ptolemaic Egypt which dates to the 2nd century BC. It shows a pair of yoked oxen driving the wheel via a sakia gear, which is here for the first time attested, too. The Greek sakia gear system is already shown fully developed to the point that "modern Egyptian devices are virtually identical". It is assumed that the scientists of the Museum of Alexandria, at the time the most active Greek research center, may have been involved in its invention. An episode from the Alexandrian War in 48 BC tells of how Caesar's enemies employed geared water wheels to pour sea water from elevated places on the position of the trapped Romans.

Around 300 AD, the noria was finally introduced when the wooden compartments were replaced with inexpensive ceramic pots that were tied to the outside of an open-framed wheel.

Poncelet Wheel

The Poncelet wheel is a type of waterwheel invented by Jean-Victor Poncelet while working at the École d'Application in Metz. It roughly doubled the efficiency of existing undershot waterwheels through a series of detail improvements. The first Poncelet wheel was constructed in 1838, and the design quickly became common in France. Although the design was a great improvement on existing designs, further improvements in turbine design rendered the Poncelet wheel obsolete by the mid-century.

Design

Traditional undershot waterwheels consist of a series of flat blades fixed to the rim of a wheel. The blades were typically mounted so they faced straight out along the radius of the wheel. When water from the millstock flowed past the wheel, it hit the blades, and some of its momentum was transferred to the wheel. However, much of it was also reflected off the blade and lost as heat. This process was not efficient; much of the original velocity in the water remained in it, meaning that potential energy was not being captured. Typical undershot wheels were around 30% efficient.

Jean Charles de Borda was the first to directly characterize the efficiency of waterwheels by comparing the velocities of water before and after meeting the wheel. Poncelet was familiar with this work and started looking for ways to improve the design. He stated that "After having reflected on this, it seemed to me that we could fulfil this double condition by replacing the straight blades on ordinary wheels with curved or cylindrical blades, presenting their concavity to the current."

His design used curved blades positioned so the water met the blade flat to its edge instead of the side. This eliminated the "bounce" that robbed power from the typical design. The water rose up into the channel between the blades for about 15 degrees of rotation, and then drained back out after another 15 degrees, where it dropped out of the channel, over the curve of the blade, imparting further impulse. By the time it left, the water had almost no velocity left. He estimated that practical wheels would reach as high as 80% for low velocity streams, and 70% for high velocity ones that fill the buckets too quickly.

Poncelet developed the design in 1823 and built a small model in 1824 that demonstrated 72% efficiency. Several commercial models followed, including a large installation in Metz that delivered 33% more power than the traditional wheel it replaced, in spite of implementing only some of the design. He published a longer paper on the design in 1826, and a much more detailed version in 1827. The design won a Prix de Mecanique from the French Academy of Sciences, who were funding development of the waterwheel and also awarded several other designs similar awards.

Poncelet wheels became common in France and Germany, where undershot designs were common. However, the large-scale installation of steam engines and water turbines led to the Poncelet wheel falling from use.

Sagebien Wheel

The Sagebien wheel is a type of water wheel invented by Alphonse Sagebien of France, a hydrological engineer and a graduate of the École Centrale des Arts et Manufactures. It was one of the most efficient breastshot water wheel designs of its era; when working on a low head of water, the Sagebien wheel could reach efficiencies of up to 90% in real-world examples.

Design

Traditional breastshot waterwheels consist of a series of flat blades fixed to the rim of a wheel. The blades were typically mounted so they faced straight out along the radius of the wheel. Water from the millstock flowed into the wheel at a point on the side of the rotation, hitting the blades and imparting momentum on the blades. The weight of the water then pulls the wheel downward through gravity. The breastshot wheel thus extracts power from both the weight and momentum of the water. Breastshot wheels reached efficiencies of up to 50%, compared to typical undershot wheels around 30%.

During a widespread French effort to improve water wheel designs, Jean Charles de Borda discovered the efficiency was a function of the relative speed of the water entering and exiting the wheel. As the incoming speed is a function of the water source, the key to extracting more power from a water wheel was reducing the velocity of the exhaust water as much as possible. He also noticed that water entering the system was often slowed before it reached the wheel, through a variety of mechanisms. He suggested that great improvements could be made by improving these aspects of wheel design.

In a conventional breastshot design, the water would flow off the edge of the headrace, and fall onto the blades of the wheel. In comparison, the Sagebien wheel had the water enter through a channel, flowing horizontally. The blades of the wheel were angled so they entered the flowing water vertically. The inside of the wheel was open, so the water could flow up into the channel without the air pressure building up and impeding the flow. The water flowed back out again after a short time, the wheel turning perhaps 30 to 45 degrees, into the lower altitude tailrace.

Sagebien built his first wheel around 1850, and made the first full-sized version, at 6 to 7 horsepower, at a flour mill in Ronquerolles in 1851. Testing demonstrated this wheel operated at about 85% efficiency, well in advance of any design of the era. This attracted little notice at the time, in the era of the steam engine, but by 1857 he had 17 in operation and people started to take note. One test in December 1861 suggested an efficiency of one model of 103%, until a new gauge reduced this to a still-excellent 93%. By 1868, more than 60 Sagebien wheels were in use in northern France, and the French Academy of Sciences awarded him the Fourneyron Prix in 1875.

Water Turbine

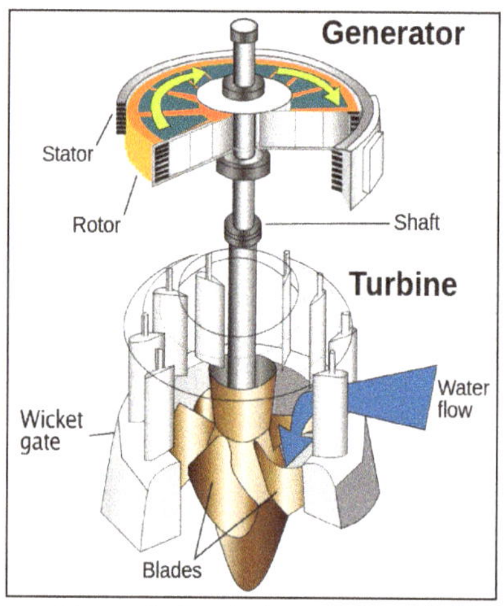

Kaplan turbine and electrical generator cut-away view.

The runner of the small water turbine.

A water turbine is a rotary machine that converts kinetic energy and potential energy of water into mechanical work.

Water turbines were developed in the 19th century and were widely used for industrial power prior to electrical grids. Now they are mostly used for electric power generation. Water turbines are mostly found in dams to generate electric power from water kinetic energy.

History

The construction of a Ganz water Turbo Generator in Budapest in 1886.

Water wheels have been used for hundreds of years for industrial power. Their main shortcoming is size, which limits the flow rate and head that can be harnessed. The migration from water wheels to modern turbines took about one hundred years. Development occurred during the Industrial revolution, using scientific principles and methods. They also made extensive use of new materials and manufacturing methods developed at the time.

Swirl

The word turbine was introduced by the French engineer Claude Burdin in the early 19th century and is derived from the Latin word for "whirling" or a "vortex". The main difference between early water turbines and water wheels is a swirl component of the water which passes energy to a spinning rotor. This additional component of motion allowed the turbine to be smaller than a water wheel of the same power. They could process more water by spinning faster and could harness much greater heads. (Later, impulse turbines were developed which didn't use swirl).

Timeline

Roman turbine mill at Chemtou, Tunisia. The tangential water inflow of the millrace made the submerged horizontal wheel in the shaft turn like a true turbine.

A Francis turbine runner, rated at nearly one million hp (750 MW), being installed at the Grand Coulee Dam, United States.

A propeller-type runner rated 28,000 hp (21 MW).

The earliest known water turbines date to the Roman Empire. Two helix-turbine mill sites of almost identical design were found at Chemtou and Testour, modern-day Tunisia, dating to the late 3rd or early 4th century AD. The horizontal water wheel with angled blades was installed at the bottom of a water-filled, circular shaft. The water from the mill-race entered the pit tangentially, creating a swirling water column which made the fully submerged wheel act like a true turbine.

Fausto Veranzio in his book Machinae Novae (1595) described a vertical axis mill with a rotor similar to that of a Francis turbine.

Johann Segner developed a reactive water turbine (Segner wheel) in the mid-18th century in Kingdom of Hungary. It had a horizontal axis and was a precursor to modern water turbines. It is a very simple machine that is still produced today for use in small hydro sites. Segner worked with Euler on some of the early mathematical theories of turbine design. In the 18th century, a Dr. Barker invented a similar reaction hydraulic turbine that became popular as a lecture-hall demonstration. The only known surviving example of this type of engine used in power production, dating from 1851, is found at Hacienda Buena Vista in Ponce, Puerto Rico.

In 1820, Jean-Victor Poncelet developed an inward-flow turbine.

In 1826, Benoît Fourneyron developed an outward-flow turbine. This was an efficient machine (~80%) that sent water through a runner with blades curved in one dimension. The stationary outlet also had curved guides.

In 1844, Uriah A. Boyden developed an outward flow turbine that improved on the performance of the Fourneyron turbine. Its runner shape was similar to that of a Francis turbine.

In 1849, James B. Francis improved the inward flow reaction turbine to over 90% efficiency. He also conducted sophisticated tests and developed engineering methods for water turbine design. The Francis turbine, named for him, is the first modern water turbine. It is still the most widely used water turbine in the world today. The Francis turbine is also called a radial flow turbine, since water flows from the outer circumference towards the centre of runner.

Inward flow water turbines have a better mechanical arrangement and all modern reaction water turbines are of this design. As the water swirls inward, it accelerates, and transfers energy to the runner. Water pressure decreases to atmospheric, or in some cases subatmospheric, as the water passes through the turbine blades and loses energy.

Around 1890, the modern fluid bearing was invented, now universally used to support heavy water turbine spindles. As of 2002, fluid bearings appear to have a mean time between failures of more than 1300 years.

Around 1913, Viktor Kaplan created the Kaplan turbine, a propeller-type machine. It was an evolution of the Francis turbine but revolutionized the ability to develop low-head hydro sites.

New Concept

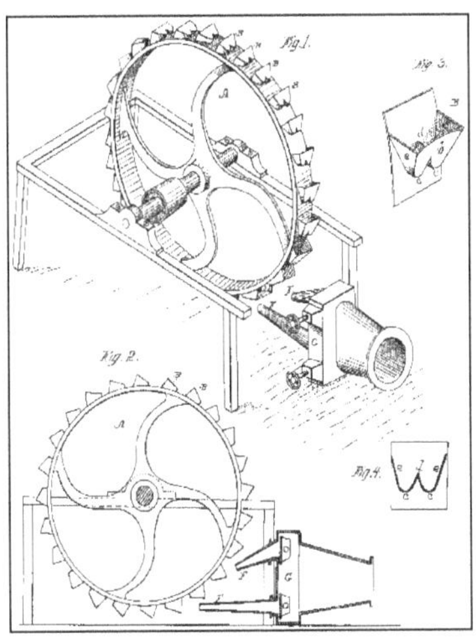

Figure from Pelton's original patent (October 1880).

All common water machines until the late 19th century (including water wheels) were basically reaction machines; water *pressure* head acted on the machine and produced work. A reaction turbine needs to fully contain the water during energy transfer.

In 1866, California millwright Samuel Knight invented a machine that took the impulse system to a new level. Inspired by the high pressure jet systems used in hydraulic mining in the gold fields, Knight developed a bucketed wheel which captured the energy of a free jet, which had converted a high head (hundreds of vertical feet in a pipe or penstock) of water to kinetic energy. This is called an impulse or tangential turbine. The water's velocity, roughly twice the velocity of the bucket periphery, does a u-turn in the bucket and drops out of the runner at low velocity.

In 1879, Lester Pelton, experimenting with a Knight Wheel, developed a Pelton wheel (double bucket design), which exhausted the water to the side, eliminating some energy loss of the Knight wheel which exhausted some water back against the center of the wheel. In about 1895, William Doble improved on Pelton's half-cylindrical bucket form with an elliptical bucket that included a cut in it to allow the jet a cleaner bucket entry. This is the modern form of the Pelton turbine which today achieves up to 92% efficiency. Pelton had been quite an effective promoter of his design and although Doble took over the Pelton company he did not change the name to Doble because it had brand name recognition.

Turgo and cross-flow turbines were later impulse designs.

Theory of Operation

Flowing water is directed on to the blades of a turbine runner, creating a force on the blades. Since the runner is spinning, the force acts through a distance (force acting through a distance is the definition of work). In this way, energy is transferred from the water flow to the turbine.

Water turbines are divided into two groups; reaction turbines and impulse turbines.

The precise shape of water turbine blades is a function of the supply pressure of water, and the type of impeller selected.

Reaction Turbines

Reaction turbines are acted on by water, which changes pressure as it moves through the turbine and gives up its energy. They must be encased to contain the water pressure (or suction), or they must be fully submerged in the water flow.

Newton's third law describes the transfer of energy for reaction turbines.

Most water turbines in use are reaction turbines and are used in low (<30 m or 100 ft) and medium (30–300 m or 100–1,000 ft) head applications. In reaction turbine pressure drop occurs in both fixed and moving blades. It is largely used in dam and large power plants.

Impulse Turbines

Impulse turbines change the velocity of a water jet. The jet pushes on the turbine's curved blades which changes the direction of the flow. The resulting change in momentum (impulse) causes a force on the turbine blades. Since the turbine is spinning, the force acts through a distance (work) and the diverted water flow is left with diminished energy. An impulse turbine is one which the pressure of the fluid flowing over the rotor blades is constant and all the work output is due to the change in kinetic energy of the fluid.

Prior to hitting the turbine blades, the water's pressure (potential energy) is converted to kinetic energy by a nozzle and focused on the turbine. No pressure change occurs at the turbine blades, and the turbine doesn't require a housing for operation.

Newton's second law describes the transfer of energy for impulse turbines.

Impulse turbines are often used in very high (>300m/1000 ft) head applications.

Power

The power available in a stream of water is;

$$P = \eta \cdot \rho \cdot g \cdot h \cdot \dot{q}$$

where:

- P = power (J/s or watts)
- η = turbine efficiency
- ρ = density of water (kg/m^3)
- g = acceleration of gravity (9.81 m/s^2)
- h = head (m). For still water, this is the difference in height between the inlet and outlet surfaces. Moving water has an additional component added to account for the kinetic energy of the flow. The total head equals the *pressure head* plus *velocity head.*
- $\dot{q}$ = flow rate (m^3/s)

Pumped-storage Hydroelectricity

Some water turbines are designed for pumped-storage hydroelectricity. They can reverse flow and operate as a pump to fill a high reservoir during off-peak electrical hours, and then revert to a water turbine for power generation during peak electrical demand. This type of turbine is usually a Deriaz or Francis turbine in design.

Efficiency

Large modern water turbines operate at mechanical efficiencies greater than 90%.

Types of Water Turbines

Various types of water turbine runners. From left to right: Pelton wheel, two types of Francis turbine and Kaplan turbine.

Reaction turbines:

- VLH turbine
- Francis turbine
- Kaplan turbine

- Tyson turbine
- Gorlov helical turbine

Impulse turbine

- Water wheel
- Pelton wheel
- Turgo turbine
- Cross-flow turbine (also known as the Bánki-Michell turbine, or Ossberger turbine)
- Jonval turbine
- Reverse overshot water-wheel
- Screw turbine
- Barkh Turbine

Design and Application

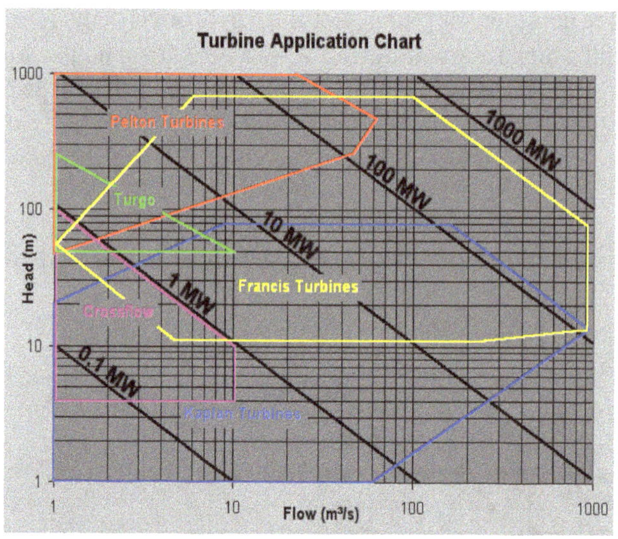

Turbine selection is based on the available water head, and less so on the available flow rate. In general, impulse turbines are used for high head sites, and reaction turbines are used for low head sites. Kaplan turbines with adjustable blade pitch are well-adapted to wide ranges of flow or head conditions, since their peak efficiency can be achieved over a wide range of flow conditions.

Small turbines (mostly under 10 MW) may have horizontal shafts, and even fairly large bulb-type turbines up to 100 MW or so may be horizontal. Very large Francis and Kaplan machines usually have vertical shafts because this makes best use of the available head, and makes installation of a generator more economical. Pelton wheels may be either vertical or horizontal shaft machines because the size of the machine is so much less than the available head. Some impulse turbines use multiple jets per runner to balance shaft thrust. This also allows for the use of a smaller turbine runner, which can decrease costs and mechanical losses.

Typical Range of Heads

• Water wheel	0.2 < H < 4 (H = head in m)
• Screw turbine	1 < H < 10
• VLH turbine	1.5 < H < 4.5
• Kaplan turbine	20 < H < 40
• Francis turbine	40 < H < 600
• Pelton wheel	50 < H < 1300
• Turgo turbine	50 < H < 250

Specific Speed

The specific speed n_s of a turbine characterizes the turbine's shape in a way that is not related to its size. This allows a new turbine design to be scaled from an existing design of known performance. The specific speed is also the main criteria for matching a specific hydro site with the correct turbine type. The specific speed is the speed with which the turbine turns for a particular discharge Q, with unit head and thereby is able to produce unit power.

Affinity Laws

Affinity laws allow the output of a turbine to be predicted based on model tests. A miniature replica of a proposed design, about one foot (0.3 m) in diameter, can be tested and the laboratory measurements applied to the final application with high confidence. Affinity laws are derived by requiring similitude between the test model and the application.

Flow through the turbine is controlled either by a large valve or by wicket gates arranged around the outside of the turbine runner. Differential head and flow can be plotted for a number of different values of gate opening, producing a hill diagram used to show the efficiency of the turbine at varying conditions.

Runaway Speed

The runaway speed of a water turbine is its speed at full flow, and no shaft load. The turbine will be designed to survive the mechanical forces of this speed. The manufacturer will supply the runaway speed rating.

Control Systems

Operation of a flyball governor to control speeds of a water turbine.

Different designs of governors have been used since the mid-19th century to control the speeds of the water turbines. A variety of flyball systems, or first-generation governors, were used during the first 100 years of water turbine speed controls. In early flyball systems, the flyball component countered by a spring acted directly to the valve of the turbine or the wicket gate to control the amount of water that enters the turbines. Newer systems with mechanical governors started around 1880. An early mechanical governors is a servomechanism that comprises a series of gears that use the turbine's speed to drive the flyball and turbine's power to drive the control mechanism. The mechanical governors were continued to be enhanced in power amplification through the use of gears and the dynamic behavior. By 1930, the mechanical governors had many parameters that could be set on the feedback system for precise controls. In the later part of the twentieth century, electronic governors and digital systems started to replace the mechanical governors. In the electronic governors, also known as second-generation governors, the flyball was replaced by rotational speed sensor but the controls were still done through analog systems. In the modern systems, also known as third-generation governors, the controls are performed digitally by algorithms that are programmed to the computer of the governor.

Turbine Blade Materials

Given that the turbine blades in a water turbine are constantly exposed to water and dynamic forces, they need to have high corrosion resistance and strength. The most common material used in overlays on carbon steel runners in water turbines are austenitic steel alloys that have 17% to 20% chromium to increase stability of the film which improves aqueous corrosion resistance. The chromium content in these steel alloys exceed the minimum of 12% chromium required to exhibit some atmospheric corrosion resistance. Having a higher chromium concentration in the steel alloys allows for a much longer lifespan of the turbine blades. Currently, the blades are made of martensitic stainless steels which have high strength compared to austenitic stainless steels by a factor of 2. Besides corrosion resistance and strength as the criteria for material selection, weld-ability and density of the turbine blade. Greater weld-ability allows for easier repair of the turbine blades. This also allows for higher weld quality which results in a better repair. Selecting a material with low density is important to achieve higher efficiency because the lighter blades rotate more easily. The most common material used in Kaplan Turbine blades are stainless steel alloys (SS). The different alloys used are SS(16Cr-5Ni), SS(13Cr-4Ni), SS(13Cr-1Ni). The martensitic stainless steel alloys have high strength, thinner sections than standard carbon steel, and reduced mass that enhances the hydrodynamic flow conditions and efficiency of the water turbine. The SS(13Cr-4Ni) has been shown to have improved erosion resistance at all angles of attack through the process of laser hardening. It is important to minimize erosion in order to maintain high efficiencies because erosion negatively impacts the hydraulic profile of the blades which reduces the relative ease to rotate.

Maintenance

Turbines are designed to run for decades with very little maintenance of the main elements; overhaul intervals are on the order of several years. Maintenance of the runners and parts exposed to water include removal, inspection, and repair of worn parts.

Normal wear and tear includes pitting corrosion from cavitation, fatigue cracking, and abrasion from suspended solids in the water. Steel elements are repaired by welding, usually with stainless steel rods. Damaged areas are cut or ground out, then welded back up to their original or an improved profile. Old turbine runners may have a significant amount of stainless steel added this way by the end of their lifetime. Elaborate welding procedures may be used to achieve the highest quality repairs.

A Francis turbine at the end of its life showing pitting corrosion, fatigue cracking and a catastrophic failure. Earlier repair jobs that used stainless steel weld rods are visible.

Other elements requiring inspection and repair during overhauls include bearings, packing box and shaft sleeves, servomotors, cooling systems for the bearings and generator coils, seal rings, wicket gate linkage elements and all surfaces.

Environmental Impact

Water turbines are generally considered a clean power producer, as the turbine causes essentially no change to the water. They use a renewable energy source and are designed to operate for decades. They produce significant amounts of the world's electrical supply.

Historically there have also been negative consequences, mostly associated with the dams normally required for power production. Dams alter the natural ecology of rivers, potentially killing fish, stopping migrations, and disrupting peoples' livelihoods. For example, American Indian tribes in the Pacific Northwest had livelihoods built around salmon fishing, but aggressive dam-building destroyed their way of life. Dams also cause less obvious, but potentially serious consequences, including increased evaporation of water (especially in arid regions), buildup of silt behind the dam, and changes to water temperature and flow patterns. In the United States, it is now illegal to block the migration of fish, for example the white sturgeon in North America, so fish ladders must be provided by dam builders.

Francis Turbine

The Francis turbine is a type of water turbine that was developed by James B. Francis in Lowell, Massachusetts. It is an inward-flow reaction turbine that combines radial and axial flow concepts.

Francis turbines are the most common water turbine in use today. They operate in a water head from 40 to 600 m (130 to 2,000 ft) and are primarily used for electrical power production. The electric generators which most often use this type of turbine have a power output which generally ranges just a few kilowatts up to 800 MW, though mini-hydro installations may be lower. Penstock (input pipes) diameters are between 3 and 33 feet (0.91 and 10.06 metres). The speed range of the turbine is from 75 to 1000 rpm. Wicket gates around the outside of the turbine's rotating runner control the rate of water flow through the turbine for different power production rates. Francis turbines are almost always mounted with the shaft vertical to isolate water from the generator. This also facilitates installation and maintenance.

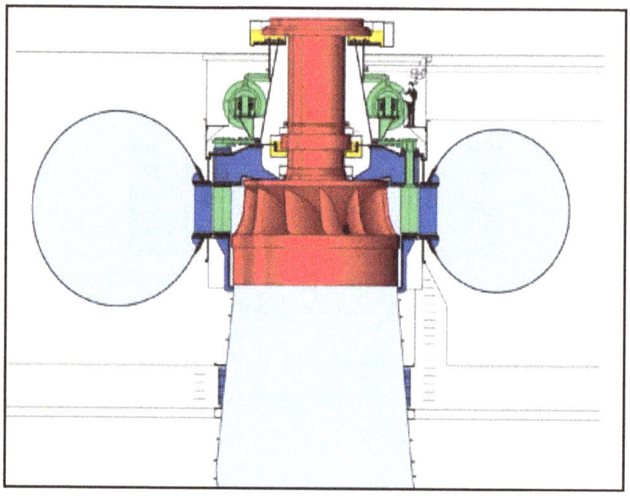

Side-view cutaway of a vertical Francis turbine. Here water enters horizontally in a spiral shaped pipe (spiral case) wrapped around the outside of the turbine's rotating *runner* and exits vertically down through the center of the turbine.

Development

Pawtucket Gatehouse in Lowell, Massachusetts; site of the first Francis turbine.

Water wheels of different types have been used historically for over 1000 years to power mills of all types, but they were relatively inefficient. Nineteenth-century efficiency improvements of water turbines allowed them to replace nearly all water wheel applications and compete with steam engines wherever water power was available. After electric generators were developed in the late 1800s turbines were a natural source of generator power where potential hydro-power sources existed.

Francis Runner, Grand Coulee Dam.

In 1826 Benoit Fourneyron developed a high efficiency (80%) outward-flow water turbine. Water was directed tangentially through the turbine runner, causing it to spin. Jean-Victor Poncelet designed an inward-flow turbine in about 1820 that used the same principles. S. B. Howd obtained a US patent in 1838 for a similar design.

In 1848 James B. Francis, while working as head engineer of the Locks and Canals company in the water wheel-powered textile factory city of Lowell, Massachusetts, improved on these designs to create more efficient turbines. He applied scientific principles and testing methods to produce a very efficient turbine design. More importantly, his mathematical and graphical calculation methods improved turbine design and engineering. His analytical methods allowed confident design of high efficiency turbines to precisely match a site's water flow and pressure (water head).

Components

A Francis turbine consists of the following main parts:

Spiral casing: The spiral casing around the runner of the turbine is known as the volute casing or scroll case. Throughout its length, it has numerous openings at regular intervals to allow the working fluid to impinge on the blades of the runner. These openings convert the pressure energy of the fluid into momentum energy just before the fluid impinges on the blades. This maintains a constant flow rate despite the fact that numerous openings have been provided for the fluid to enter the blades, as the cross-sectional area of this casing decreases uniformly along the circumference.

Guide or stay vanes: The primary function of the guide or stay vanes is to convert the pressure energy of the fluid into the momentum energy. It also serves to direct the flow at design angles to the runner blades.

Runner blades:Runner blades are the heart of any turbine. These are the centers where the fluid strikes and the tangential force of the impact causes the shaft of the turbine to rotate, producing torque. Close attention in design of blade angles at inlet and outlet is necessary, as these are major parameters affecting power production.

Draft tube: The draft tube is a conduit which connects the runner exit to the tail race where the water is discharged from the turbine. Its primary function is to reduce the velocity of discharged water to minimize the loss of kinetic energy at the outlet. This permits the turbine to be set above the tail water without appreciable drop of available head.

Theory of Operation

Three Gorges Dam Francis turbine runner.

The Francis turbine is a type of reaction turbine, a category of turbine in which the working fluid comes to the turbine under immense pressure and the energy is extracted by the turbine blades from the working fluid. A part of the energy is given up by the fluid because of pressure changes occurring in the blades of the turbine, quantified by the expression of Degree of reaction, while the remaining part of the energy is extracted by the volute casing of the turbine. At the exit, water acts on the spinning cup-shaped runner features, leaving at low velocity and low swirl with very little kinetic or potential energy left. The turbine's exit tube is shaped to help decelerate the water flow and recover the pressure.

Francis Turbine (exterior view) attached to a generator.

Cut-away view, with wicket gates (yellow) at minimum flow setting.

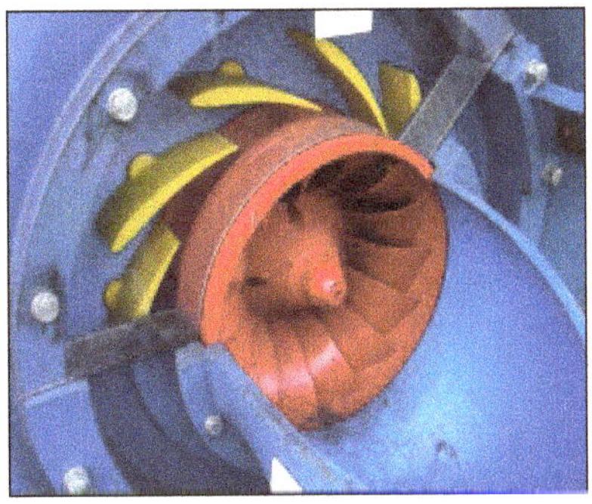

Cut-away view, with wicket gates (yellow) at full flow setting.

Blade Efficiency

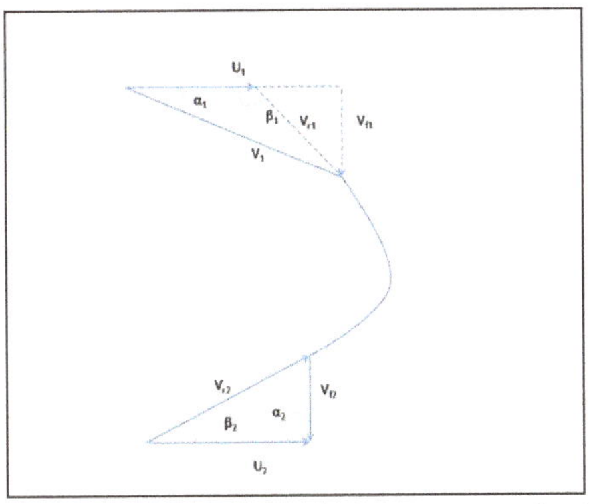

Ideal velocity diagram, illustrating that in ideal cases the whirl component of outlet velocity is zero and the flow is completely axial.

Usually the flow velocity (velocity perpendicular to the tangential direction) remains constant throughout, i.e. $V_{f1}=V_{f2}$ and is equal to that at the inlet to the draft tube. Using Euler turbine equation, $E/m=e=V_{w1}U_1$, where e is the energy transfer to the rotor per unit mass of the fluid. From the inlet velocity triangle,

$$V_{w1} = V_{f1}\cot\alpha_1$$

and

$$U_1 = V_{f1}(\cot\alpha_1 + \cot\beta_1),$$

Therefore,

$$e = V_{f1}^2 \cot\alpha_1(\cot\alpha_1 + \cot\beta_1).$$

The loss of kinetic energy per unit mass becomes $V_{f2}^2/2$.

Therefore, neglecting friction, the blade efficiency becomes

$$\eta_b = \frac{e}{(e+V_{f2}^2/2)},$$

i.e.

$$\eta_b = \frac{2V_{f1}^2(\cot\alpha_1(\cot\alpha_1 + \cot\beta_1))}{V_{f2}^2 + 2V_{f1}^2(\cot\alpha_1(\cot\alpha_1 + \cot\beta_1))}.$$

Degree of Reaction

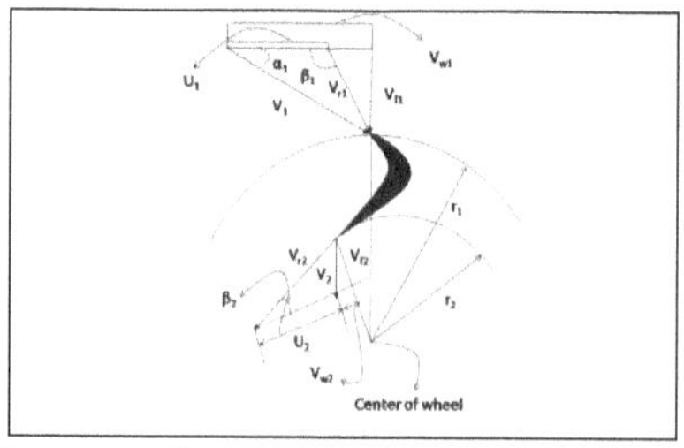

Actual velocity diagram, illustrating that the whirl component of the outlet velocity is non-zero.

Degree of reaction can be defined as the ratio of pressure energy change in the blades to total energy change of the fluid. This means that it is a ratio indicating the fraction of total change in fluid pressure energy occurring in the blades of the turbine. Rest of the changes occur in the stator blades of the turbines and the volute casing as it has a varying cross-sectional area. For example, if the degree of reaction is given to be 50%, that means that half of the total energy change of the fluid is taking place in the rotor blades and the other half is occurring in the stator blades. If the degree of reaction is zero it means that the energy changes due to the rotor blades is zero, leading to a different turbine design called Pelton Turbine.

$R=e-1/2(V_1^2-V_{f2}^2)/e$

Now, putting the value of 'e' from above and using ($V_1^2-V_{f2}^2=V_{f1}^2\cot\alpha_2$ as $V_{f2}=V_{f1}$)

$R=1-(\cot\alpha_1/2(\cot\alpha_1+\cot\beta_1))$

Application

Francis Inlet Scroll, Grand Coulee Dam.

Francis turbines may be designed for a wide range of heads and flows. This, along with their high efficiency, has made them the most widely used turbine in the world. Francis type units cover a head range from 40 to 600 m (130 to 2,000 ft), and their connected generator output power varies from just a few kilowatts up to 800 MW. Large Francis turbines are individually designed for each site to operate with the given water supply and water head at the highest possible efficiency, typically over 90%.

Small Swiss-made Francis turbine.

In contrast to the Pelton turbine, the Francis turbine operates at its best completely filled with water at all times. The turbine and the outlet channel may be placed lower than the lake or sea level outside, reducing the tendency for cavitation.

In addition to electrical production, they may also be used for pumped storage, where a reservoir is filled by the turbine (acting as a pump) driven by the generator acting as a large electrical motor during periods of low power demand, and then reversed and used to generate power during peak demand. These pump storage reservoirs act as large energy storage sources to store "excess" electrical energy in the form of water in elevated reservoirs. This is one of a few methods that temporary excess electrical capacity can be stored for later utilization.

Kaplan Turbine

A Bonneville Dam Kaplan turbine after 61 years of service.

The Kaplan turbine is a propeller-type water turbine which has adjustable blades. It was developed in 1913 by Austrian professor Viktor Kaplan, who combined automatically adjusted propeller blades with automatically adjusted wicket gates to achieve efficiency over a wide range of flow and water level.

The Kaplan turbine was an evolution of the Francis turbine. Its invention allowed efficient power production in low-head applications that was not possible with Francis turbines. The head ranges from 10–70 metres and the output from 5 to 200 MW. Runner diameters are between 2 and 11 metres. Turbines rotate at a constant rate, which varies from facility to facility. That rate ranges from as low as 69.2 rpm (Bonneville North Powerhouse, Washington U.S.) to 429 rpm. The Kaplan turbine installation believed to generate the most power from its nominal head of 34.65 m is as of 2013 the Tocoma Dam Power Plant (Venezuela) Kaplan turbine generating 235 MW with each of ten 4.8 m diameter runners.

Kaplan turbines are now widely used throughout world in high-flow, low-head power production.

On this Kaplan runner the pivots at the base of the blade are visible; these allow the angle of the blades to be changed while running. The hub contains hydraulic cylinders for adjusting the angle.

Development

Viktor Kaplan living in Brno, Czech Republic, obtained his first patent for an adjustable blade propeller turbine in 1912. But the development of a commercially successful machine would take another decade. Kaplan struggled with cavitation problems, and in 1922 abandoned his research for health reasons.

In 1919 Kaplan installed a demonstration unit at Poděbrady, Czechoslovakia. In 1922 Voith introduced an 1100 HP (about 800 kW) Kaplan turbine for use mainly on rivers. In 1924 an 8 MW unit went on line at Lilla Edet, Sweden. This marked the commercial success and widespread acceptance of Kaplan turbines.

Theory of Operation

The Kaplan turbine is an outward flow reaction turbine, which means that the working fluid changes pressure as it moves through the turbine and gives up its energy. Power is recovered from both the hydrostatic head and from the kinetic energy of the flowing water. The design combines features of radial and axial turbines.

The inlet is a scroll-shaped tube that wraps around the turbine's wicket gate. Water is directed tangentially through the wicket gate and spirals on to a propeller shaped runner, causing it to spin.

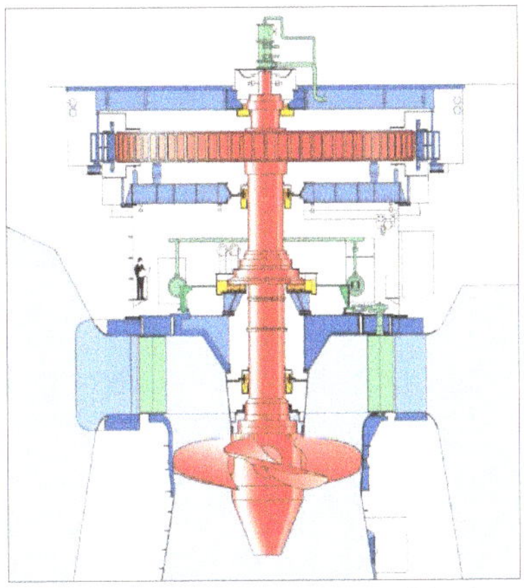

Vertical Kaplan Turbine (courtesy Voith-Siemens).

The outlet is a specially shaped draft tube that helps decelerate the water and recover kinetic energy.

The turbine does not need to be at the lowest point of water flow as long as the draft tube remains full of water. A higher turbine location, however, increases the suction that is imparted on the turbine blades by the draft tube. The resulting pressure drop may lead to cavitation.

Variable geometry of the wicket gate and turbine blades allow efficient operation for a range of flow conditions. Kaplan turbine efficiencies are typically over 90%, but may be lower in very low head applications.

Current areas of research include CFD driven efficiency improvements and new designs that raise survival rates of fish passing through.

Because the propeller blades are rotated on high-pressure hydraulic oil bearings, a critical element of Kaplan design is to maintain a positive seal to prevent emission of oil into the waterway. Discharge of oil into rivers is not desirable because of the waste of resources and resulting ecological damage.

Applications

Viktor Kaplan Turbine Technisches Museum Wien.

Kaplan turbines are widely used throughout the world for electrical power production. They cover the lowest head hydro sites and are especially suited for high flow conditions.

Inexpensive micro turbines on the Kaplan turbine model are manufactured for individual power production designed for 3 m of head which can work with as little as 0.3 m of head at a highly reduced performance provided sufficient water flow.

Large Kaplan turbines are individually designed for each site to operate at the highest possible efficiency, typically over 90%. They are very expensive to design, manufacture and install, but operate for decades.

They have recently found a new home in offshore wave energy generation.

Variations

The Kaplan turbine is the most widely used of the propeller-type turbines, but several other variations exist:

- *Propeller turbines* have non-adjustable propeller vanes. They are used in where the range of flow / power is not large. Commercial products exist for producing several hundred watts from only a few feet of head. Larger propeller turbines produce more than 100 MW. At the La Grande-1 generating station in northern Quebec, 12 propeller turbines generate 1368 MW.
- *Bulb* or *tubular turbines* are designed into the water delivery tube. A large bulb is centered in the water pipe which holds the generator, wicket gate and runner. Tubular turbines are a fully axial design, whereas Kaplan turbines have a radial wicket gate.
- *Pit turbines* are bulb turbines with a gear box. This allows for a smaller generator and bulb.
- *Straflo turbines* are axial turbines with the generator outside of the water channel, connected to the periphery of the runner.
- *S-turbines* eliminate the need for a bulb housing by placing the generator outside of the water channel. This is accomplished with a jog in the water channel and a shaft connecting the runner and generator.
- The *VLH turbine* an open flow, very low head "kaplan" turbine slanted at an angle to the water flow. It has a large diameter >3.55m, is low speed using a directly connected shaft mounted permanent magnet alternator with electronic power regulation and is very fish friendly (<5% mortality).
- *Tyson turbines* are a fixed propeller turbine designed to be immersed in a fast flowing river, either permanently anchored in the river bed, or attached to a boat or barge.

model of a bulb or tubular turbines.

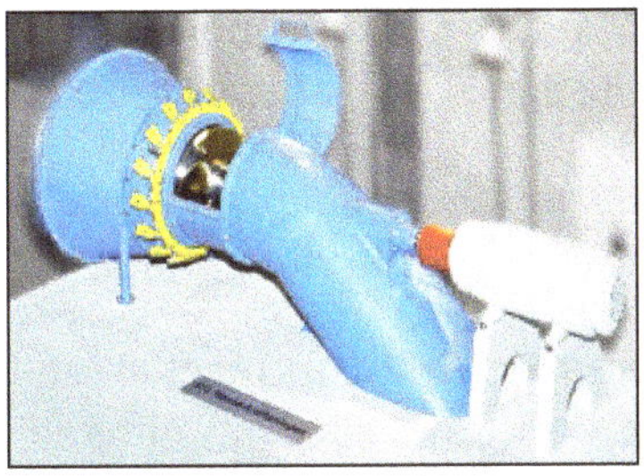

model of a S-turbine.

Cross-flow Turbine

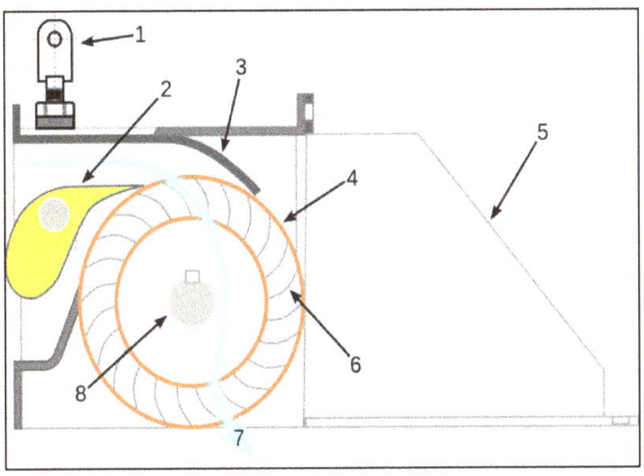

Diagram of a Cross-flow turbine
1 — air-venting valve
2 — distributor
3 — turbine casing (all thick grey)
4 — runner
5 — removable rear casing
6 — blades
7 — water flow
8 — shaft

A cross-flow turbine, Bánki-Michell turbine, or Ossberger turbine is a water turbine developed by the Australian Anthony Michell, the Hungarian Donát Bánki and the German Fritz Ossberger. Michell obtained patents for his turbine design in 1903, and the manufacturing company Weymouth made it for many years. Ossberger's first patent was granted in 1933 ("Free Jet Turbine" 1922, Imperial Patent No. 361593 and the "Cross Flow Turbine" 1933, Imperial Patent No. 615445), and he manufactured this turbine as a standard product. Today, the company founded by Ossberger is the leading manufacturer of this type of turbine.

Unlike most water turbines, which have axial or radial flows, in a cross-flow turbine the water passes through the turbine transversely, or across the turbine blades. As with a water wheel, the water is admitted at the turbine's edge. After passing to the inside of the runner, it leaves on the opposite side, going outward. Passing through the runner twice provides additional efficiency.

When the water leaves the runner, it also helps clean it of small debris and pollution. The cross-flow turbine is a low-speed machine that is well suited for locations with a low head but high flow.

Although the illustration shows one nozzle for simplicity, most practical cross-flow turbines have two, arranged so that the water flows do not interfere.

Cross-flow turbines are often constructed as two turbines of different capacity that share the same shaft. The turbine wheels are the same diameter, but different lengths to handle different volumes at the same pressure. The subdivided wheels are usually built with volumes in ratios of 1:2. The subdivided regulating unit, the guide vane system in the turbine's upstream section, provides flexible operation, with 33, 66 or 100% output, depending on the flow. Low operating costs are obtained with the turbine's relatively simple construction.

Details of Design

Ossberger turbine section.

The turbine consists of a cylindrical water wheel or runner with a horizontal shaft, composed of numerous blades (up to 37), arranged radially and tangentially. The blade's edges are sharpened to reduce resistance to the flow of water. A blade is made in a part-circular cross-section (pipe cut over its whole length). The ends of the blades are welded to disks to form a cage like a hamster cage and are sometimes called "squirrel cage turbines"; instead of the bars, the turbine has the trough-shaped steel blades.

The water flows first from the outside of the turbine to its inside. The regulating unit, shaped like a vane or tongue, varies the cross-section of the flow. The water jet is directed towards the cylindrical runner by nozzle. The water enters the runner at an angle of about 45/120 degrees,transmitting some of the water's kinetic energy to the active cylindrical blades.

Ossberger turbine runner.

The regulating device controls the flow based on the power needed, and the available water. The ratio is that (0–100%) of the water is admitted to 0-100%×30/4 blades. Water admission to the two nozzles is throttled by two shaped guide vanes. These divide and direct the flow so that the water enters the runner smoothly for any width of opening. The guide vanes should seal to the edges of the turbine casing so that when the water is low, they can shut off the water supply. The guide vanes therefore act as the valves between the penstock and turbine. Both guide vanes can be set by control levers, to which an automatic or manual control may be connected.

The turbine geometry (nozzle-runner-shaft) assures that the water jet is effective. The water acts on the runner twice, but most of the power is transferred on the first pass, when the water enters the runner. Only ⅓ of the power is transferred to the runner when the water is leaving the turbine.

The water flows through the blade channels in two directions: outside to inside, and inside to outside. Most turbines are run with two jets, arranged so two water jets in the runner will not affect each other. It is, however, essential that the turbine, head and turbine speed are harmonised.

The cross-flow turbine is of the impulse type, so the pressure remains constant at the runner.

Advantages

The peak efficiency of a cross-flow turbine is somewhat less than a Kaplan, Francis or Pelton turbine. However, the cross-flow turbine has a flat efficiency curve under varying load. With a split runner and turbine chamber, the turbine maintains its efficiency while the flow and load vary from 1/6 to the maximum.

Since it has a low price, and good regulation, cross-flow turbines are mostly used in mini and micro hydropower units of less than two thousand kW and with heads less than 200 m.

Particularly with small run-of-the-river plants, the flat efficiency curve yields better annual performance than other turbine systems, as small rivers' water is usually lower in some months. The efficiency of a turbine determines whether electricity is produced during the periods when rivers have low flows. If the turbines used have high peak efficiencies, but behave poorly at partial load, less annual performance is obtained than with turbines that have a flat efficiency curve.

Due to its excellent behaviour with partial loads, the cross-flow turbine is well-suited to unattended electricity production. Its simple construction makes it easier to maintain than other turbine types; only two bearings must be maintained, and there are only three rotating elements. The mechanical system is simple, so repairs can be performed by local mechanics.

Another advantage is that it can often clean itself. As the water leaves the runner, leaves, grass etc. will not remain in the runner, preventing losses. Therefore, although the turbine's efficiency is somewhat lower, it is more reliable than other types. No runner cleaning is normally necessary, e.g. by flow inversion or variations of the speed. Other turbine types are clogged more easily, and consequently face power losses despite higher nominal efficiencies.

Gorlov Helical Turbine

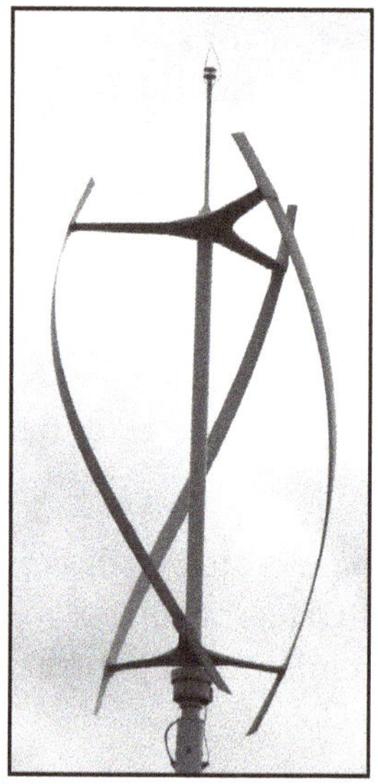

quietrevolution QR5 wind turbine.

The Gorlov helical turbine (GHT) is a water turbine evolved from the Darrieus turbine design by altering it to have helical blades/foils. It was patented in a series of patents from September 19, 1995 to July 3, 2001 and won 2001 ASME Thomas A. Edison Patent Award. GHT was invented by Professor Alexander M. Gorlov of Northeastern University.

A.M.Gorlov with his turbine.

The physical principles of the GHT work are the same as for its main prototype, the Darrieus turbine, and for the family of similar Vertical axis wind turbines which includes also Turby wind turbine Quietrevolution wind turbine Urban Green Energy. GHT, turby and quietrevolution solved pulsatory torque issues by using the helical twist of the blades.

Fluid Performance

The term "foil" is used to describe the shape of the blade cross-section at a given point, with no distinction for the type of fluid, (thus referring to either an "airfoil" or "hydrofoil"). In the helical design, the blades curve around the axis, which has the effect of evenly distributing the foil sections throughout the rotation cycle, so there is always a foil section at every possible angle of attack. In this way, the sum of the lift and drag forces on each blade do not change abruptly with rotation angle. The turbine generates a smoother torque curve, so there is much less vibration and noise than in the Darrieus design. It also minimizes peak stresses in the structure and materials, and facilitates self-starting of the turbine. In testing environments the GHT has been observed to have up to 35% efficiency in energy capture reported by several groups. "Among the other vertical-axis turbine systems, the Davis Hydro turbine, the EnCurrent turbine, and the Gorlov Helical turbine have all undergone scale testing at laboratory or sea. Overall, these technologies represent the current norm of tidal current development."

Turbine Axis-orientation

The main difference between the Gorlov helical turbine and conventional turbines is the orientation of the axis in relation to current flow. The GHT is a vertical-axis turbine which means the axis is positioned perpendicular to current flow, whereas traditional turbines are horizontal-axis turbines which means the axis is positioned parallel to the flow of the current. Fluid flows, such as wind, will naturally change direction, however they will still remain parallel to the ground. So in all vertical-axis turbines, the flow remains perpendicular to the axis, regardless of the flow direction, and the turbines always rotate in the same direction. This is one of the main advantages of vertical-axis turbines.

If the direction of the water flow is fixed, then the Gorlov turbine axis could be vertical or horizontal, the only requirement is orthogonality to the flow.

Airfoil / Hydrofoil

The GHT operates under a lift-based concept. The foil sections on the GHT are symmetrical, both top-to-bottom and also from the leading-to-trailing edge. The GHT can actually spin equally well in either direction. The GHT works under the same principle as the Darrieus turbine; that is, it relies upon the movement of the foils in order to change the apparent direction of the flow relative to the foils, and thus change the (apparent) "angle of attack" of the foil.

Environmental Issues

A GHT is proposed for low-head micro hydro installations, when construction of a dam is undesirable. The GHT is an example of damless hydro technology. The technology may potentially offer cost and environmental benefits over dam-based micro-hydro systems.

Some advantages of damless hydro are that it eliminates the potential for failure of a dam, which improves public safety. It also eliminates the initial cost of dam engineering, construction and maintenance, reduces the environmental and ecological complications, and potentially simplifies the regulatory issues put into law specifically to mitigate the problems with dams.

In general, a major ecological issue with hydropower installations is their actual and perceived risk to aquatic life. It is claimed that a GHT spins slowly enough that fish can see it soon enough to swim around it. From preliminary tests in 2001, it was claimed that if a fish swims between the slowly moving turbine blades, the fish will not be harmed. Also it would be difficult for a fish to become lodged or stuck in the turbine, because the open spaces between the blades are larger than even the largest fish living in a small river. A fish also would not be tumbled around in a vortex, because the GHT does not create a lot of turbulence, so small objects would be harmlessly swept through with the current.

How it Works

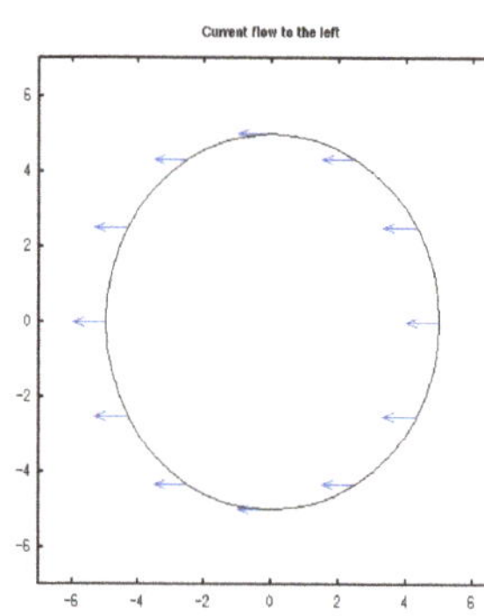

Current flow to the left.

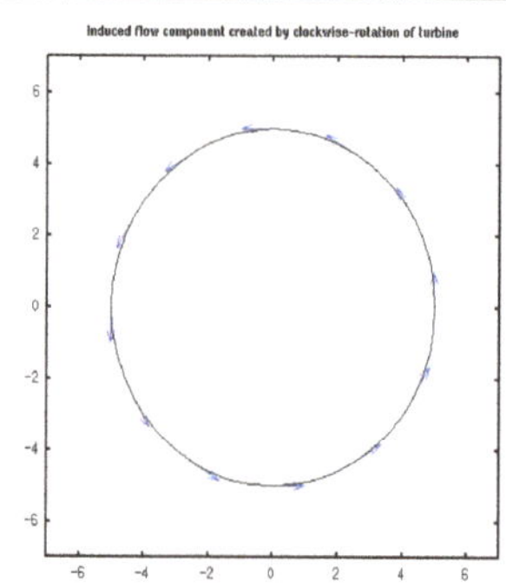

Induced flow component created by clockwise rotation of turbine.

In this example the direction of the fluid flow is to the left.

As the turbine rotates, in this case in a clockwise direction, the motion of the foil through the fluid changes the apparent velocity and angle of attack (speed and direction) of the fluid with respect to the frame of reference of the foil. The combined effect of these two flow components (i.e. the vector sum), yields the net total "Apparent flow velocity" as shown in the next figure.

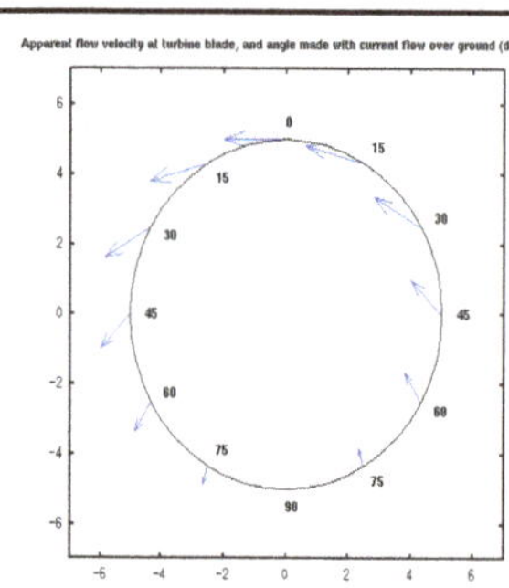

Apparent flow velocity of turbine blade, and angle made with current flow over ground (in degrees).

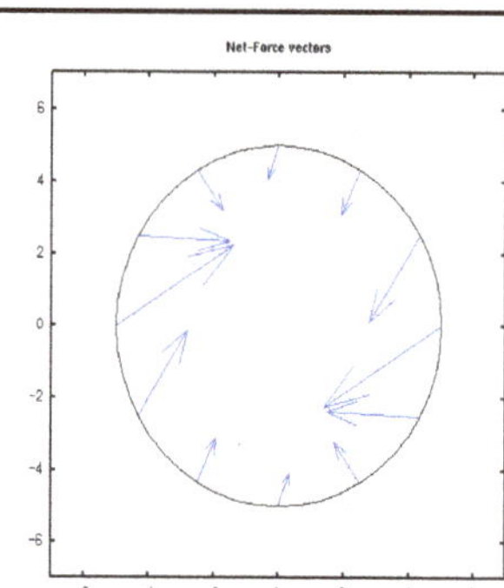

Net force vectors.

The action of this apparent flow on each foil section generates both a lift and drag force, the sum of which is shown in the figure above titled "Net force vectors". Each of these net force vectors can be split into two orthogonal vectors: a radial component and a tangential component, shown here as "Normal force" and "Axial force" respectively. The normal forces are opposed by the rigidity of the turbine structure and do not impart any rotational force or energy to the turbine. The remaining force component propels the turbine in the clockwise direction, and it is from this torque that energy can be harvested.

[With regards to the figure above left "Apparent flow velocity...", Lucid Energy Technologies, rights holder to the patent to the Gorlov Helical Turbine, notes that this diagram, with no apparent velocity at an azimuth angle of 180 degrees (blade at its point in rotation where it is instantaneously moving in downstream direction), may be subject to misinterpretation. This is because a zero apparently flow velocity could occur only at a tip speed ratio of unity (i.e. TSR=1, where the current flow induced by rotation equals the current flow). The GHT generally operates at a TSR substantially greater than unity.]

(The diagrams "Net Force Vectors" and "Normal Force Vectors" are partially incorrect. The downwind segments should show the vectors outside the circles. Otherwise there would be no net sideways loading on the turbine.) M Koester 2015.

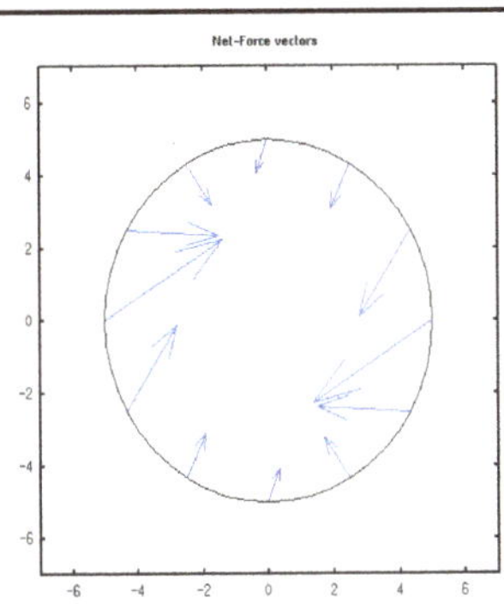

Normal force vectors.

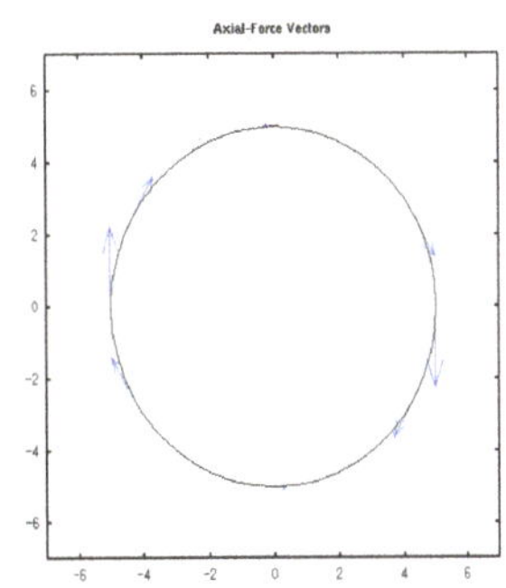

Axial force vectors.

Commercial Use

Helical turbines in water stream generate mechanical power independent on direction of the water flow. Then electric generators assembled upon the common shaft transfer the power into electricity for the commercial use.

Chain of Horizontal Gorlov Turbines. TideGen by Ocean Renewable Power Company – possibility for shallow waters. (US Patents 5,451,137, Sep. 1995 and 5,642,984, Jul. 1997)

Chain of Horizontal Gorlov turbines being installed in Cobscook Bay, Maine, USA.

Tidal Power Station with Gorlov Helical Turbines before deployment in the ocean. USA, Cobscook Bay, Maine, September, 2012.

Chain of Horizontal Gorlov turbines being installed in South Korea - general view.

Chain of Horizontal Gorlov turbines being installed in South Korea - close view.

Gorlov Helical Turbines in South Korea, 1997-1998. Installation in shallow waters.

Pelton Wheel

Old Pelton wheel from Walchensee Hydroelectric Power Station, Germany.

Assembly of a Pelton wheel at Walchensee Hydroelectric Power Station, Germany.

The Pelton wheel is an impulse type water turbine. It was invented by Lester Allan Pelton in the

1870s. The Pelton wheel extracts energy from the impulse of moving water, as opposed to water's dead weight like the traditional overshot water wheel. Many variations of impulse turbines existed prior to Pelton's design, but they were less efficient than Pelton's design. Water leaving those wheels typically still had high speed, carrying away much of the dynamic energy brought to the wheels. Pelton's paddle geometry was designed so that when the rim ran at half the speed of the water jet, the water left the wheel with very little speed; thus his design extracted almost all of the water's impulse energy—which allowed for a very efficient turbine.

Function

Nozzles direct forceful, high-speed streams of water against a rotary series of spoon-shaped buckets, also known as impulse blades, which are mounted around the circumferential rim of a drive wheel—also called a runner. As the water jet impinges upon the contoured bucket-blades, the direction of water velocity is changed to follow the contours of the bucket. Water impulse energy exerts torque on the bucket-and-wheel system, spinning the wheel; the water stream itself does a "u-turn" and exits at the outer sides of the bucket, decelerated to a low velocity. In the process, the water jet's momentum is transferred to the wheel and hence to a turbine. Thus, "impulse" energy does work on the turbine. For maximum power and efficiency, the wheel and turbine system is designed such that the water jet velocity is twice the velocity of the rotating buckets. A very small percentage of the water jet's original kinetic energy will remain in the water, which causes the bucket to be emptied at the same rate it is filled, and thereby allows the high-pressure input flow to continue uninterrupted and without waste of energy. Typically two buckets are mounted side-by-side on the wheel, which permits splitting the water jet into two equal streams. This balances the side-load forces on the wheel and helps to ensure smooth, efficient transfer of momentum of the fluid jet of water to the turbine wheel.

Because water and most liquids are nearly incompressible, almost all of the available energy is extracted in the first stage of the hydraulic turbine. Therefore, Pelton wheels have only one turbine stage, unlike gas turbines that operate with compressible fluid. It is used for generating electricity.

Applications

Pelton wheels are the preferred turbine for hydro-power, when the available water source has relatively high hydraulic head at low flow rates, where the Pelton wheel geometry is most suitable. Pelton wheels are made in all sizes. There exist multi-ton Pelton wheels mounted on vertical oil pad bearings in hydroelectric plants. The largest units can be over 400 megawatts. The smallest Pelton wheels are only a few inches across, and can be used to tap power from mountain streams having flows of a few gallons per minute. Some of these systems use household plumbing fixtures for water delivery. These small units are recommended for use with 30 metres (100 ft) or more of head, in order to generate significant power levels. Depending on water flow and design, Pelton wheels operate best with heads from 15–1,800 metres (50–5,910 ft), although there is no theoretical limit.

Design Rules

The specific speed η_s parameter is independent of a particular turbine's.

Compared to other turbine designs, the relatively low specific speed of the Pelton wheel, implies that the geometry is inherently a "low gear" design. Thus it is most suitable to being fed by a hydro source with a low ratio of flow to pressure, (meaning relatively low flow and/or relatively high pressure).

The specific speed is the main criterion for matching a specific hydro-electric site with the optimal turbine type. It also allows a new turbine design to be scaled from an existing design of known performance.

$\eta_s = n\sqrt{P} / \sqrt{\rho}(gH)^{5/4}$ (dimensioned parameter),

where:

- n = Frequency of rotation (rpm)
- P = Power (W)
- H = Water head (m)
- ρ = Density (kg/m^3)

The formula implies that the Pelton turbine is *geared* most suitably for applications with relatively high hydraulic head H, due to the 5/4 exponent being greater than unity, and given the characteristically low specific speed of the Pelton.

Turbine Physics and Derivation

Energy and Initial Jet Velocity

In the ideal (frictionless) case, all of the hydraulic potential energy ($E_p = mgh$) is converted into kinetic energy ($E_k = mv^2/2$). Equating these two equations and solving for the initial jet velocity (V_i) indicates that the theoretical (maximum) jet velocity is $V_i = \sqrt{(2gh)}$. For simplicity, assume that all of the velocity vectors are parallel to each other. Defining the velocity of the wheel runner as: (u), then as the jet approaches the runner, the initial jet velocity relative to the runner is: ($V_i - u$). The initial jet velocity of jet is V_i .

Final Jet Velocity

Assuming that the jet velocity is higher than the runner velocity, if the water is not to become backed-up in runner, then due to conservation of mass, the mass entering the runner must equal the mass leaving the runner. The fluid is assumed to be incompressible (an accurate assumption for most liquids). Also it is assumed that the cross-sectional area of the jet is constant. The jet *speed* remains constant relative to the runner. So as the jet recedes from the runner, the jet velocity relative to the runner is: $-(V_i - u) = -V_i + u$. In the standard reference frame (relative to the earth), the final velocity is then: $V_f = (-V_i + u) + u = -V_i + 2u$.

Optimal Wheel Speed

We know that the ideal runner speed will cause all of the kinetic energy in the jet to be transferred to the wheel. In this case the final jet velocity must be zero. If we let $-V_i + 2u = 0$, then the optimal runner speed will be $u = V_i/2$, or half the initial jet velocity.

Torque

By Newton's second and third laws, the force F imposed by the jet on the runner is equal but opposite to the rate of momentum change of the fluid, so:

$$F = -m(V_f - V_i) = -\rho Q[(-V_i + 2u) - V_i] = -\rho Q(-2V_i + 2u) = 2\rho Q(V_i - u)$$

where (ρ) is the density and (Q) is the volume rate of flow of fluid. If (D) is the wheel diameter, the torque on the runner is: $T = F(D/2) = \rho QD(V_i - u)$. The torque is at a maximum when the runner is stopped (i.e. when $u = 0$, $T = \rho QDV_i$). When the speed of the runner is equal to the initial jet velocity, the torque is zero (i.e. when $u = V_i$, then $T = 0$). On a plot of torque versus runner speed, the torque curve is straight between these two points, ($0, \rho QDV_i$) and (V_i, 0).

Power

The power $P = Fu = T\omega$, where ω is the angular velocity of the wheel. Substituting for F, we have $P = 2\rho Q(V_i - u)u$. To find the runner speed at maximum power, take the derivative of P with respect to u and set it equal to zero, $[dP/du = 2\rho Q(V_i - 2u)]$. Maximum power occurs when $u = V_i/2$. $P_{max} = \rho QV_i^2/2$. Substituting the initial jet power $V_i = \sqrt{(2gh)}$, this simplifies to $P_{max} = \rho ghQ$. This quantity exactly equals the kinetic power of the jet, so in this ideal case, the efficiency is 100%, since all the energy in the jet is converted to shaft output.

Efficiency

A wheel power divided by the initial jet power, is the turbine efficiency, $\eta = 4u(V_i - u)/V_i^2$. It is zero for $u = 0$ and for $u = V_i$. As the equations indicate, when a real Pelton wheel is working close to maximum efficiency, the fluid flows off the wheel with very little residual velocity. In theory, the energy efficiency varies only with the efficiency of the nozzle and wheel, and does not vary with hydraulic head. The term "efficiency" can refer to: Hydraulic, Mechanical, Volumetric, Wheel, or overall efficiency.

System Components

The conduit bringing high-pressure water to the impulse wheel is called the penstock. Originally the penstock was the name of the valve, but the term has been extended to include all of the fluid supply hydraulics. Penstock is now used as a general term for a water passage and control that is under pressure, whether it supplies an impulse turbine or not.

Turgo Turbine

The Turgo turbine is an impulse water turbine designed for medium head applications. Operational Turgo Turbines achieve efficiencies of about 87%. In factory and lab tests Turgo Turbines perform with efficiencies of up to 90%. It works with net heads between 15 and 300 m.

Developed in 1919 by Gilkes as a modification of the Pelton wheel, the Turgo has some advantages over Francis and Pelton designs for certain applications.

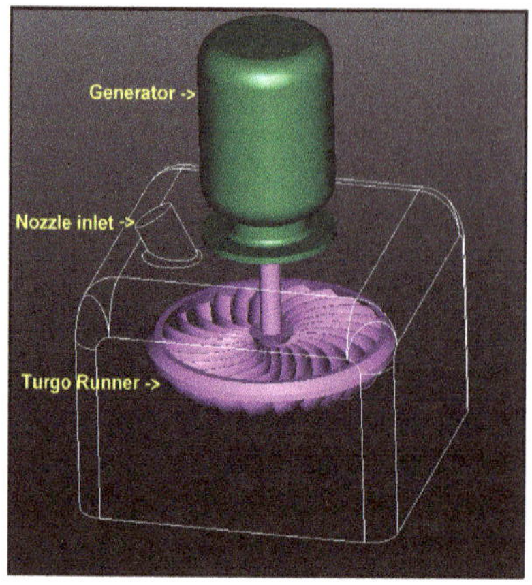

Turgo turbine and generator.

First, the runner is less expensive to make than a Pelton wheel. Second, it doesn't need an airtight housing like the Francis. Third, it has higher specific speed and can handle a greater flow than the same diameter Pelton wheel, leading to reduced generator and installation cost.

Turgos operate in a head range where the Francis and Pelton overlap. While many large Turgo installations exist, they are also popular for small hydro where low cost is very important.

Like all turbines with nozzles, blockage by debris must be prevented for effective operation.

At Milford sound, New Zealand.

Theory of Operation

The Turgo turbine is an impulse type turbine; water does not change pressure as it moves through the turbine blades. The water's potential energy is converted to kinetic energy with a nozzle. The high speed water jet is then directed on the turbine blades which deflect and reverse the flow. The resulting impulse spins the turbine runner, imparting energy to the turbine shaft. Water exits with very little energy. Turgo runners are extremely efficient.

A Turgo runner looks like a Pelton runner split in half. For the same power, the Turgo runner is one half the diameter of the Pelton runner, and so twice the specific speed. The Turgo can handle a greater water flow than the Pelton because exiting water doesn't interfere with adjacent buckets.

The specific speed of Turgo runners is between the Francis and Pelton. Single or multiple nozzles can be used. Increasing the number of jets increases the specific speed of the runner by the square root of the number of jets (four jets yield twice the specific speed of one jet on the same turbine).

References

- Adriana de Miranda (2007). Water architecture in the lands of Syria: the water-wheels. L'Erma di Bretschneider. pp. 38–9. ISBN 88-8265-433-8
- Langdon, John (2004), Mills in the Medieval Economy: England, 1300-1540, Oxford: Oxford University Press, ISBN 0-19-926558-5
- McErlean, Thomas; Crothers, Norman (2007), Harnessing the Tides: The Early Medieval Tide Mills at Nendrum Monastery, Strangford Lough, Belfast: Stationery Office Books, ISBN 978-0-337-08877-3
- C Rossi; F Russo; F Russo (2009). "Ancient Engineers' Inventions: Precursors of the Present". Springer. ISBN 904812252X
- *Nevell, Mike; Walker (2001). Portland Basin and the archaeology of the Canal Warehouse. Tameside Metropolitan Borough with University of Manchester Archaeological Unit. ISBN 1-871324-25-4
- Örjan Wikander (2008). "Chapter 6: Sources of Energy and Exploitation of Power". In John Peter Oleson. The Oxford Handbook of Engineering and Technology in the Classical World. Oxford University Press. pp. 141–2. ISBN 0-19-518731-8
- de Crespigny, Rafe (2007), A Biographical Dictionary of Later Han to the Three Kingdoms (23-220 AD), Leiden: Koninklijke Brill, ISBN 90-04-15605-4
- Thomson, Ross (2009). Structures of Change in the Mechanical Age: Technological Invention in the United States 1790-1865. Baltimore, MD: The Johns Hopkins University Press. p. 34. ISBN 978-0-8018-9141-0
- Rynne, Colin (2000), "Waterpower in Medieval Ireland", in Squatriti, Paolo, Working with Water in Medieval Europe, Technology and Change in History, 3, Leiden: Brill, pp. 1–50, ISBN 90-04-10680-4

Chapter 5

Ocean Energy

Ocean energy encompasses all forms of renewable energy carried by ocean waves, salinity and tides. The energy which is generated by using the kinetic energy from the rise and fall of tides is known as tidal energy. This chapter closely examines the key concepts of ocean energy to provide an extensive understanding of the subject.

Ocean energy (also referred to as marine energy) refers to the energy carried by ocean waves, tides, salinity, and ocean temperature differences. The movement of water in the world's oceans creates a vast store of kinetic energy, or energy in motion. This energy can be harnessed to generate electricity to power homes, transport and industries.

Types of Ocean Energy

The ocean can produce two types of energy:

- Thermal energy from the sun's heat.
- Mechanical energy from the tides and waves.

Oceans cover more than 70% of Earth's surface, making them the world's largest solar energy collectors. The sun's heat warms the surface water a lot more than the deep ocean water, and this temperature difference creates thermal energy.

Ocean thermal energy is used for many applications, including electricity generation. Ocean thermal energy conversion (OTEC), uses ocean temperature differences from the surface to depths lower than 1,000 meters, to extract energy. A temperature difference of only 20 °C can yield usable energy.

There are three types of electricity conversion systems:

- Closed-cycle systems use the ocean's warm surface water to vaporise a working fluid, which has a low-boiling point, such as ammonia. The vapour expands and turns a turbine. The turbine then activates a generator to produce electricity.
- Open-cycle systems actually boil the seawater by operating at low pressures. This produces steam that passes through a turbine/generator.
- Hybrid systems combine both closed-cycle and open-cycle systems.

Ocean mechanical energy is quite different from ocean thermal energy. Even though the sun affects all ocean activity, tides are driven primarily by the gravitational pull of the moon, and waves are driven primarily by the winds. As a result, tides and waves are intermittent sources of energy. A barrage (dam) is typically used to convert tidal energy into electricity by forcing the water through turbines, activating a generator. Tidal energy is extremely site specific requires mean tidal differences greater than 4 meters and also favourable topographical conditions, such as estuaries and bays in order to bring down costs of dams.

Wave energy describes energy generated from the power of waves near their surface. Several different types of wave energy conversion devices extract power from motion of waves. These include single point absorbers. The power take off device within these systems converts the motion of the waves into electrical energy. Wave power can also be captured through oscillating water columns, which trap waves in a column and change the air pressure in the upper portion which drives a turbine. Wave power can also be captured through an overtopping device, which traps waves in a floating pool, which is then released through turbines to generate power.

Tidal Energy

Tides are the periodic motion of the waters of the sea due to the inter-attractive forces between the celestial bodies. Tides are very long-period waves that move through the oceans in response to the forces exerted by the moon and sun. Tide and current are not the same. Tide is the vertical rise and fall of the water and tidal current is the horizontal flow. In simple words, the tide rises and falls, the tidal current floods and ebbs. The principal of tidal forces are generated by the Moon and Sun. The Moon is the main tide-generating body. Due to its greater distance, the Sun's effect is only 46 per cent of the Moon's.

Energy from the Tides

There are three types of tides: diurnal, semidiurnal and mixed. Tidal Energy is one of the new and evolving technologies, which is commercially not viable and still in Research & Development (R&D) stage. Tidal energy is inexhaustible and can be considered as a renewable energy source. It is an advantage because it is less vulnerable to climate change; while the other sources are all vulnerable to the random changes in climate. The review given by the Energy Technology Support Unit (ETSU) on the Tidal Stream Energy was the initial attempt to estimate the energy from tidal stream resources in the UK. The points marked by the ESTU were later studied and modified in 2001 in a document submitted to the UK Department of Trade and Industry (DTI) by Binnie, Black and Veatch. Most of the existing technology used for tidal energy conversion is from the wind power industry. Researchers have predicted that UK has is capable to produce over 20% of its electrical needs from its tidal resources. It is also a fact that the studies carried out so far in predicting the energy that can be extracted from tides, has only focused on the past and present availability of the energy. But it is also important to consider and address the effects of exploiting the renewable energy sources for energy extraction. There has to be an understanding among the developers as to when and where to stop the energy extraction so that there is minimum or no disturbance caused to the regular natural phenomenon.

Tidal Energy around the World

The necessity to reduce CO_2 emissions and gradual increase in cost of fossil fuel has resulted in a significantly increased use of tidal energy. Today, tidal energy around the world is increasingly being considered as a potential source of renewable energy. Extreme tides are found in many locations across the globe. Some of them are: the Pentland Firth, Scotland; the Severn estuary; the Aleutians; the fjords of Norway; the Philippines; the Straits of Messina, Italy; the Bosporus, Turkey; the English Channel; Indonesia, and the straits of Alaska and British Columbia.

The first major hydroelectric plant was put to operation in 1967 that used the energy of the tides to generate electricity. It produced about 540,000 kW of electricity. Studies have shown that the European territorial waters have 106 locations for extracting tidal energy that would provide electricity of 48 TW per year. It is estimated around 50,000 MW of installed capacity being achievable along the coasts of British Columbia alone. There are greater predictions of extracting energy of about 90,000 MW off the North West coast of Russia and about 20,000 MW at the inlet or Mezen river and White Sea. There are also estimations along the West coast of India having potential to generate 8,000MW.

Table gives the highest available tidal levels in some of the regions that have the potential to establish tidal power stations. Tidal power plants have already been set up at some of these places and some are still in the planning phase. The main characteristics of four large-scale tidal power plants that were constructed after World War II and currently exist are given in table.

Table: Highest tides of the global ocean.

Site	Country	Tidal elevation (m)
Bay of Fundy	Canada	16.2
Severn Estuary	England	14.5
Port of Ganville	France	14.7
La Rance	France	13.5
Puerto Rio Gallegos	Argentina	13.3
Bay of Mezen (White Sea)	Russia	10
Penzhinskaya Guba (Sea of Okhotsk)	Russia	13.4

Table: Existing large tidal power plants.

Site	Country	Bay area (km^2)	Avg. tide (m)	Installed Power (MW)
La Rance	France	22	8.55	240
Kislaya Guba	Russia	1.1	2.3	0.4
Annapolis	Canada	15	6.4	18
Jiangxia	China	1.4	5.08	3.9

Methods of Tidal Energy Extraction

Different methods have been suggested by authors for the extraction of tidal energy. However the basic principle behind the methods remains same. However, there are two primary methods to extract energy from the tides:

- Estuaries into which large amounts of ocean water flows in due to high tidal range are captured behind barrages and the turbines are rotated by utilizing the potential energy of the stored water.
- The kinetic energy of moving water can be used to extract energy similar to the principle of extraction of wind energy.

Both methods that are mentioned above have been suggested and followed and each has its own advantages and disadvantages. It may also be possible to employ pumping strategies for barrages to obtain better efficiency and to match electricity demand better. The devices that are used in the energy generation vary in size, shape and specifications. ISSC has classified the devices into three types:

- Tidal barrages that store tidal flow and generate power through discharge.
- Tidal fences which block a passage and extract energy in either or both directions of tidal flow.
- Tidal current devices which are fixed or moored within a tidal stream.

Tidal Barrage

Tidal barrage is a structure generally built across the mouth of the estuary through which the water flows in and out of the basin. The tidal barrage has sluice gate that allows the flow of water in and out of the basin. The water flows into the bay during high tide and the water is retained by closing the sluice gates at the beginning of low tide. The barrage gates are controlled by knowing the tidal range of the location and operating it at right times of the tidal cycle. There are turbines located at the sluice gates which produce electricity when the gates are opened during the low tide. Figure shows the Plan view of a hypothetical tidal barrage. Even though the barrage method has high theoretical efficiency, only one large scale tidal barrage has been constructed at La Rance, France.

The advantage of using barrage to method to generate electricity in comparision with fossil fuels is that it reduces the greenhouse effects, to provide a better environment. La Rance tidal power plant, France is an example for barrage method.On the top of the barrage there is a four-lane highway that cuts 35 km of distance between the towns of Saint Malo and Dinard representing.

Top view of La Rance tidal power plant barrage (750m in length).

Tidal Stream Energy

Tidal power was mainly focused on harnessing the tidal flow and generating the energy by means of potential storage rather than through tidal stream. Tidal stream technologies have made massive progress towards commercialization in the last decade. Extensive research is being carried out in UK waters related to tidal stream energy. UK has a target at achieve 20% of its electricity requirement through ocean resources by 2020. About 40 energy converting machines are being developed and prorotypes are being tested in the labs and in waters of UK. Since the tidal stream energy is still a emerging technology, it has no standardizations, but variety of devices are being developed to make use of the water flow to extract electricity. However, the efficiency of each of the devices has to be flawlessly examined by extensive testing to choose the appropriate device for a particular location.

Calculation of Tidal Energy

The total tidal energy is the energy due to the tidal stream (kinetic energy) and the energy due to release of the stored water in the basin (potential energy). It is also a fact that the increase in tidal variation or the tidal stream energy results in increase of energy extraction to a large extent.

Kinetic Energy

The kinetic energy of the stream flow flowing across the cross section with a velocity is given by:

$$P=\frac{1}{2}\rho A V^3 C_P$$

- ρ is the density of sea water (kg/m³).
- C_P is the power coefficient.
- A is the area of cross section of the channel (m²).
- V is current velocity (m/s).

The power output or the efficiency of the turbine "ξ" depends on the design of the turbine. The power output for a turbine from these kinetic systems can be obtained by the following equation:

$$P = \frac{\xi A \rho V^3}{2}$$

- ξ is turbine efficiency.
- P is power generated (in watts).
- ρ is density of the water (seawater is 1025 kg/m³).
- A is sweep area of the turbine (in m²).
- V is velocity of the flow.

Potential Energy

The potential energy is mainly dependent on the tidal prism of the basin. Potential energy obtained due to the stored water can be calculated as:

$$E = \frac{1}{2} A \rho g h^2$$

- h is the vertical tidal range.
- A is the horizontal area of the barrage basin.
- ρ is the density of water = 1025 kg per cubic meter (seawater varies between 1021 and 1030 kg/m³).
- g is the acceleration due to the Earth's gravity = 9.81 m/s².

From previous equation, it can be seen that the potential energy varies with square of tidal range. So, a barrage should be placed in such a location where it is possible to achieve maximum storage head. Black and Veatch suggest that the ideal water depths to achieve the best possible power output at few potential sites around the UK range between 25 and 40 m and the recommend diameter of the rotor to range between 10 m and 20 m. The inlets that are between islands having large basin area are considered to have a greater amplification effect because of the reduction in the throat area and the water depth relative to the surroundings, producing a venturi effect. This accelerates the water as it is forced through a channel with a smaller cross-sectional area.

Tidal Energy Devices

There is wide range of tidal energy devices used for energy extraction based on site conditions. The details of some of them are given in table.

Table: Tidal energy extraction devices.

Turbine	Power Output	Description	Image
Seagen, Marine Current Turbines Ltd. UK	1 MW	Twin two bladed rotors, sheath mounted, horizontal axis tidal turbine.	
The Blue Concept, Hammerfest Strom, Norway	1 MW	Three bladed pile mounted horizontal axis tidal turbine.	
THG, Tidal Hydraulic Generator	3 MW	Array of four bladed horizontal axis turbines attached to gravity mounted frame.	
TidEL SMD Hydrovision, UK	1 MW	Twin two bladed tethered rotor, able to yaw into the current.	
Stingray, Engineering Business Ltd., UK	500 kW	Oscillating Hydroplane.	
Blue energy, Blue energy Inc., Canada	500 kW	Four bladed, moored, surface piercing, vertical axis turbine.	

Recommended Sites for Installation of Tidal Power Plants

Some preliminary standards are given by Couch and Bryden to identify sites that are suitable for the development of a tidal energy extraction. The most important variables generally considered are:

- The local water depth: Existing device technology concepts are generally limited to operational water depths of 25–45 metres.
- The location of the nearest exploitable grid connection: For an immature industry, the economics of tidal energy extraction require easy access to a nearby grid connection with spare capacity; otherwise the capital cost cannot be viably recouped across the life of the project.
- An energetic and persistent resource: Large mean spring and neap tide velocities are highly desirable. Some sites have the added advantage of minimizing the low velocity periods of the tidal cycle as the local dynamics ensure that the tidal flow reverses through the slack period at an accelerated rate. The sites that the developers are interested to extract energy tend to have peak spring tidal velocities of 3+ m/s.

If these three primary criteria are met, a site is considered to have solid potential for future development. The majority of coastal locations can be rejected out of hand by consideration of just these three variables.

The English Channel, the Arctic Ocean, The Gulf of Mexico, The Amazon, The Straits of Magellan and Taiwan are some of the possible locations for location tidal devices. Table shows some potential sites for tidal power installations. There are estimates that the energy that can be globally extracted is around 1800 TWh/year. But, it has to be taken care that the effects on environment, economic and social constraints have to addressed. It is suggested that 10% of the extraction of energy can be considered as guideline for harnessing renewable energy resources. The estimated construction costs for existing and proposed tidal barrages are given in table.

Table: Estimated construction costs for existing and proposed tidal barrages.

Barrage	Country	Capacity (MW)	Power Generation (GWh)	Construction costs (Million USD)	Construction costs per kW (USD/kW)
Operating					
La Rance	France	240	540	817	340
Sihwa Lake	Korea	254	552	298	117
Proposed/planned					
Gulf of Kutch	India	50	100	162	324
Wyre Barrage	UK	61.4	131	328	534

Garorim Bay	Korea	520	950	800	154
Mersey Barrage	UK	700	1340	5741	820
Incheon	Korea	1320	2410	3772	286
Dalupiri Blue	Philippines	2200	4000	3034	138
Severn Barrage	UK	8640	15600	36085	418
Penzhina Bay	Russia	87000	200000	328066	377

Table: Some potential sites for tidal power installations.

Site	Country	Bay area (Km²)	Avg. Tide	Potential Power (MW)
Passamaquoddy	USA	300	5.5	400
Cook Inlet	USA	3100	4.35	Up to 18000
Mezen	Russia	2640	5.66	15000
Tugur	Russia	1080	5.38	6790
Severn	UK	490	8.3	6000
Mersey	UK	60	8.4	700
San Jose	Argentina	780	6.0	7000
Carolim Bay	Korea	90	4.7	480
Secure	Australia	130	8.4	570
Walcott	Australia	260	8.4	1750

Technology and economics, however, dictate that only those areas where currents can exceed 2m/s might be suitable for exploitation.

Advantages and Disadvantages of Tidal Power

Advantages

- Tidal power is a renewable and sustainable energy resource. It reduces dependence upon fossil fuels.
- It produces no liquid or solid pollution. It has little visual impact.
- Tidal power exists on a worldwide scale from deep ocean waters.
- Tidally driven coastal currents provide an energy density four times greater than air, which means that a 15m diameter turbine will generate as much energy generated by a 60mdiameter windmill.
- Tidal currents are both predictable and reliable, a feature which gives them an advantage over both wind and solar systems. Power outputs can be accurately calculated far in advance, allowing for easy integration with existing electricity grids.

Disadvantages

- High cost of construction, installation and generation.
- Barrages can disrupt natural migratory routes for marine animals and normal boating pathways.
- Turbines can kill up to 15% of fish in area, although technology has advanced, the turbines have to move slow enough not to kill many.
- Flooding and ecological changes.
- Research is still in initial stages.

Types of Tidal Turbines

Several types for turbines are considered suitable for harnessing tidal energy. Their main classification is based on the form of energy to be harnessed namely (a) potential energy by barrage based tidal energy convertor, and (b) kinetic energy by tidal stream based hydrokinetic turbines.

Turbines for Barrage based Tidal Energy Convertors

The barrage based tidal energy convertors are similar to conventional hydroelectric power plants in construction. These convertors employ axial flow turbines named bulb turbines or tubular turbines to operate at low heads and high flow rates offered by tidal barrages. The flow over turbine reverses during tide cycle. The blades of these turbines are designed to suit this flow reversal. Rotor blades of the bulb turbines look very similar to a ship's propeller. By virtue of the blade construction, these turbines spin in opposite direction as the flow reverses. The name bulb or tubular turbines is due to a large bulb at the center of the water tube that houses the turbine. The bulb holds the generator, wicket gate and runner. Wicket gate is essentially a set of guide vanes that control the flow of water over the turbine blades.

Cutaway section of a large bulb turbine showing the bulb that houses gearbox and generator.

Generally these turbines are large in size and weight which is also an important factor as its cost further add to the capital investment. One of the bulb turbine units in La Rance Tidal Energy Plant has a tip diameter of 5.35 metres and it weighs 470 tonnes. It has four blades and 24 guide vanes. Its operating head range is from 3 metre to 11 metre with peak output of 10 MW.

Tidal Stream Turbines

Oceans, covering more than 70 % of the earth, have long been appreciated as a vast renewable energy source. The energy is stored in oceans partly as thermal energy, partly as kinetic energy (waves and currents) and also in chemical and biological products. The tidal stream turbines are also known as underwater windmills. They are driven by the kinetic energy of the moving water in similar way that wind turbines use moving air. The generator is placed into a marine current that typically results when water being moved by tidal forces comes up against, or moves around, an obstacle or through a constriction such as a passage between two masses of land. A tidal generator converts the energy of tidal flows into electricity. Greater tidal variation and higher tidal current velocities can dramatically increase the potential of a site for tidal electricity generation.

Working

Tidal Stream turbines operate on the same principles that wind turbines use. A flow of water moves a set of blades creating mechanical energy which is then converted to electrical energy by the generator. Tidal stream generators (or TSGs) make use of the kinetic energy of moving water to power turbines, in a similar way to wind turbines that use wind to power turbines. Some tidal generators can be built into the structures of existing bridges, involving virtually no aesthetic problems. Tidal Stream turbine uses underwater spinning blades to turn a generator. Underwater turbines rely on tides to push water against angled blades, causing them to spin. As the blades spin, a gearbox turns an induction generator, which produces an electric current. These turbines must be able to swivel 180 degrees to accommodate the ebb and flow of tides. These turbines can be placed in natural bodies of water, such as harbors and lagoons that naturally feature fast-moving flows of water.

Construction

The technology consists of rotors mounted on steel piles (tubular steel columns) set into a socket drilled in the seabed. A key requirement for tidal stream devices is the support structure concepts to hold them in place. Currently there are three options under consideration: Gravity Structure, Piled Structure, and Floating Structure.

Gravity Structure

Gravity Structures are massive steel or concrete structures attached to the base of the units to achieve stability by their own inertia.

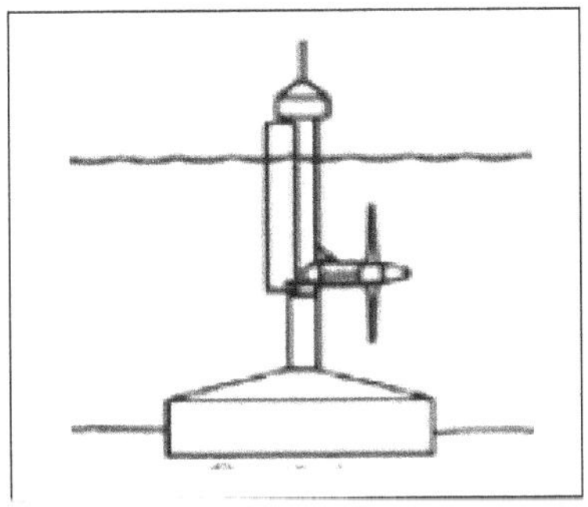

Piled Structure

Piled Structures are pinned to the seabed by one or more steel or concrete piles. The piles are fixed to the seabed by hammering if the ground conditions are sufficiently soft or by pre-drilling, positioning and grouting if the rock is harder.

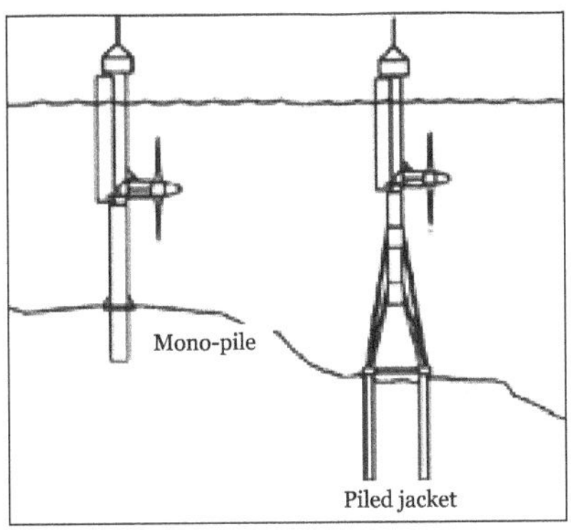

Floating Structure

The turbine unit is mounted on a downward pointing vertical column rigidly fixed to a flatboat. The flatboat is then fixed to the seabed by chains or wire ropes which hang in a centenary and may be fixed to the seabed by drag, piled or gravity anchors, depending on the seabed condition.

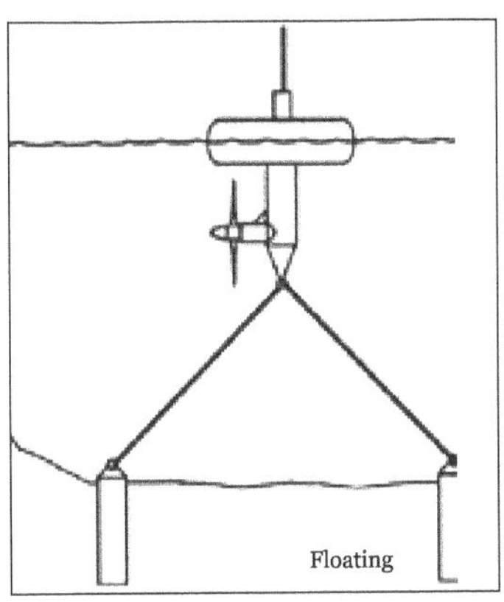

Types of Tidal Stream Turbine

Horizontal Axis Turbine

Horizontal axis turbines work much the same as a conventional wind turbine and some look very similar in design. A turbine is placed in a tidal stream which causes the turbine to rotate and produce power. A prototype of a 300 KW horizontal axis turbine was installed in kvalsund, south of Hammerfest, Norway and connected to the grid on 13 Nov. 2003. A 300 KW horizontal axis turbine, also known as sea flow, was installed by marine current turbines off the coast of Lynmouth, Dxevon, England, in 2003. A prototype project was installed in the East River between Queens and Roosevelt Island in New York City in the United States in April 2007. A prototype, called SeaGen, was installed by Marine Current Turbines in Stangford Lough in Northern Ireland in April 2008.

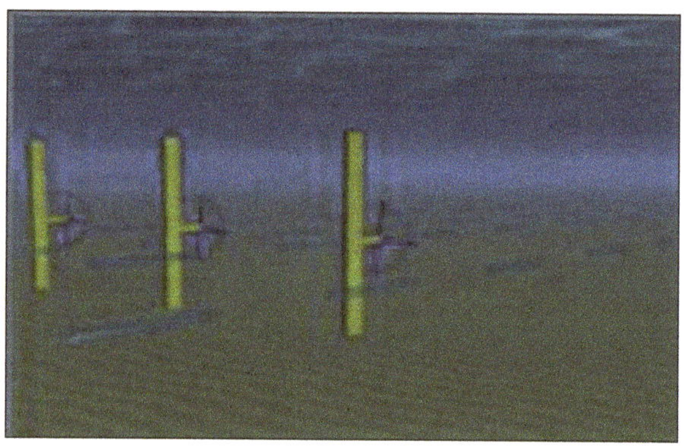

Vertical Axis Turbine

Vertical axis turbines use the same principle as the horizontal axis turbines only with a different direction of rotation. The Enermar Project developed Kobold turbine. A pilot plant of is moored in the Strait of Messina, close to the Sicilian shore in Italy, in an average sea tidal current of about 2 m/sec.

Reciprocating Devices (Oscillating Hydrofoils)

These have hydrofoils which move back and forth in a plane normal to the tidal stream, instead of rotating blades.

The oscillation motion used to produce power is due to the lift created by the tidal stream flowing in either side of the wing. During 2003, a 150 KW Stingray was tested off the Scottish coast, including a flexible control system to allow the performance of the generator to be accurately controlled and recorded over a longer period.

Venturi Effect Tidal Stream Devices

The tidal flow is directed through a duct, which concentrates the flow and produces a pressure difference. This causes a secondary fluid flow through a turbine. The resultant flow can drive a turbine directly or the induced pressure differential in the system can drive an air-turbine.

The tidal Energy installed and tested such shrouded tidal turbines on the Gold coast, Queensland in 2002. Another turbine is planned for deployment as a tidal powered desalination showcase near Brisbane Australia in October 2008. Another device, the hydro Venturi, is to be tested in San Francisco Bay.

Advantages and Disadvantages

Because water is far denser than air, spinning blades can potentially be more productive than off-shore wind turbines for the same amount of space. Another advantage, water turbines enjoy over other sources of renewable energy is a predictable tide table.

As wind turbines interrupt bird flights just as water turbines can disturb underwater life. Fishing has to be restricted in the areas of the power plant. Another issue tidal turbine technology faces during active life is chemical pollution due to necessary coatings and lubrications. Chemicals coatings are used on the turbine to control naturally occurring corrosion. Although tidal turbines are inaudible from land, another issue pertaining to the health of the surrounding ecosystem is the level of noise that the turbines produce beneath the ocean surface.

Cross Flow Turbine

Cross flow turbines are widely used on UK small hydro sites with typical power outputs from 5 kW up to 100 kW, though they can actually be up to 3 MW in size on the very largest systems. Generally speaking there are better turbine choices for these higher power outputs.

They will work on net heads from just 1.75 metres all of the way to 200 metres, though there are more appropriate turbine choices for sites with heads above 40 metres. They will work on average annual flows as low as 40 litres/second up to 5 m^3/s, though on the higher flow rates there may be better other turbine types to consider.

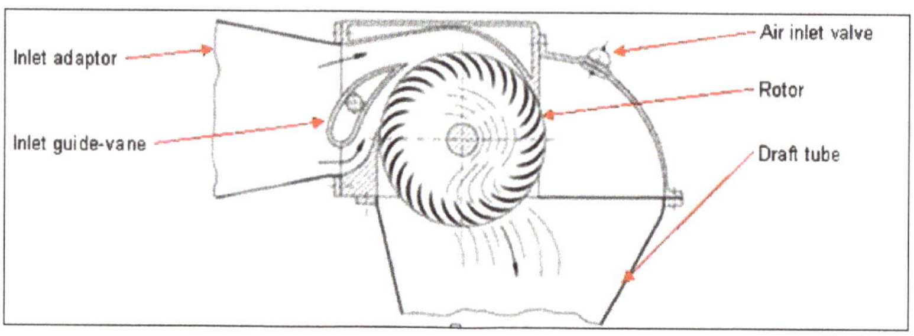

Crossflow turbine in cross-section.

Crossflow turbines gets their name from the way the water flows through, or more correctly 'across' the rotor as shown in figure (hence across flow or crossflow). The water flows over and under the inlet guide-vane which directs flow to ensure that the water hits the rotor at the correct angle for maximum efficiency. The water then flows over the upper rotor blades, producing a torque on the rotor, then through the centre of the rotor and back across the low rotor blades producing more torque on the rotor. Most of the power is extracted by the upper blades (roughly 75%) and the remaining 25% by the lower blades. Obviously the rotor is rotating, so what are the upper blades one moment will be the lower blades the next.

One of the advantages of crossflow turbines are that they are self-cleaning to a degree in that leaves etc. that could get pushed into and stuck on the upper blades are washed off by the exiting water on the lower blades. Also the centrifugal force tends to throw trapped debris outwards, further increasing the self-cleaning capabilities. This isn't such a great benefit nowadays because fine fish screens are required at the intake, which tends to exclude debris anyway.

The mains parts are shown in figures. The water enters through the inlet adaptor, also called the 'round to square adaptor' because it fits between the round water supply pipe and the rectangular inlet to the turbine. The inlet adaptor can be in a horizontal orientation or vertical to suit the site conditions.

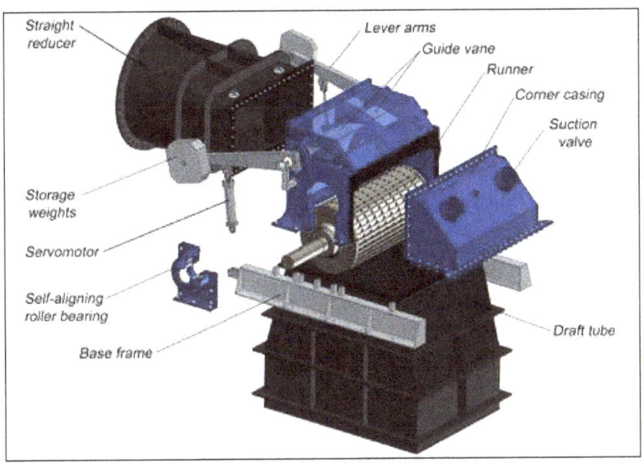

Main parts of a crossflow turbine.

Next the water comes to the inlet guide-vanes, which both regulate the flow rate through the turbine and direct the flow onto the rotor at the optimum angle for maximum efficiency. The rotor spins on a horizontal axis and looks similar to a cylinder lawn mower rotor, though is of a much heavier construction.

Crossflow turbines are impulse turbines, which means (amongst other things) that the rotor is spinning air and is not fully-flooded like in a reaction (e.g. Kaplan) turbine. The water exits the rotor and falls into the draft tube, which can be many metres long (though not normally more than 1/3 of the total net head across the system). The exiting water fills the draft tube from just below the rotor all of the way to the discharge water level. This column of water waters wants to 'fall out' of the draft tube, but cannot because the air that would need to fill void is prevented from entering quickly by the air inlet valve on the front turbine casing. This creates a negative pressure inside the turbine cavity and is called the 'suction head' across the turbine. The positive water pressure from the upstream side of the rotor is called the 'pressure head' and the sum of the pressure and suction head equals the net head across the hydro system.

It is important when setting up crossflows to adjust the air inlet valve correctly so that it allows just the right amount of air into the turbine chamber; too much and the column of water in the draft tube will fall, reducing the suction head and the efficiency of the system. Too little and the column of water in the draft tube will rise and 'drown' the rotor, which causes excessive drag and also reduces efficiency.

The advantage of using a draft tube to maximise the suction head is that the long draft tube moves the main body of the crossflow turbine upwards, away from the discharge water level and (hopefully) above the floodwater level during high flow events. If the main body of the turbine was still likely to get flooded during high flow events it would have to be built into a 'tanked' enclosure.

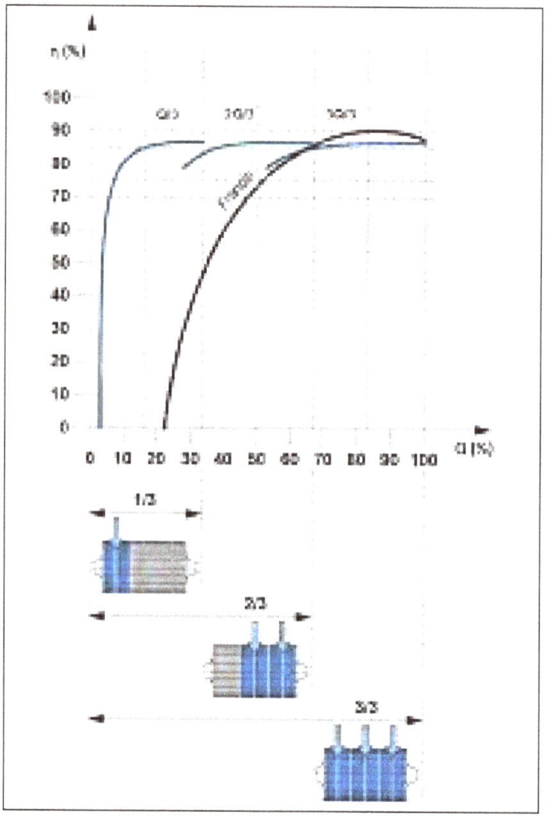

Efficiency curve for 1/3, 2/3 and 3/3 inlet guide-vanes in operation.

A typical low-head crossflow turbine installation is shown in section in figure. What is notable is the depth of the discharge sump underneath the draft tube exit. This is required to make sure that the discharge water can exit efficiently, but does mean that a very deep excavation is required below the site downstream water level during the construction phase. This can be problematic if 100% effective coffer dams cannot be built to dry-out the area due to the site layout, and is quite an expensive feature to construct.

Another major advantage of a crossflow turbine is their wide operating flow range with a high efficiency across the whole range. This is possible because (most) crossflows have two inlet guide-vanes, one 1/3 of the intake width and the second 2/3 of the width (a so-called 1/3:2/3 split). This means that during lower flow periods the 2/3 inlet guide-vane can be completely closed allowing no water through, and the turbine will operate on just the 1/3 guide-vane which effectively means that only 1/3 of the rotor is in use. If 'average' flow rates are available the 1/3 guide-vane can close and the turbine operates on just the 2/3 side, then when high flow rates are available both guide-vanes can work together. This is shown diagrammatically in figure along with the efficiency curve. Note that the peak efficiency shown of 86% is a little high: 82% is more realistic for a high quality crossflow turbine. The horizontal axis (labelled 'Q (%)') is the percentage of the maximum flow rate, so 100% is the maximum flow rate, 50% half of maximum etc. The turbine efficiency is shown on the vertical axis.

In good quality crossflows the guide-vanes fit the turbine casing with such precision that they can stop the water flow 100%. This is a useful feature because it means the rotor can be stopped fully when the turbine is shut down, though in some situations it is better to leave the rotor turning very slowly even when shut down to prevent various creepy crawlies taking up residence inside the generator if it shut down for a period during summer low flows.

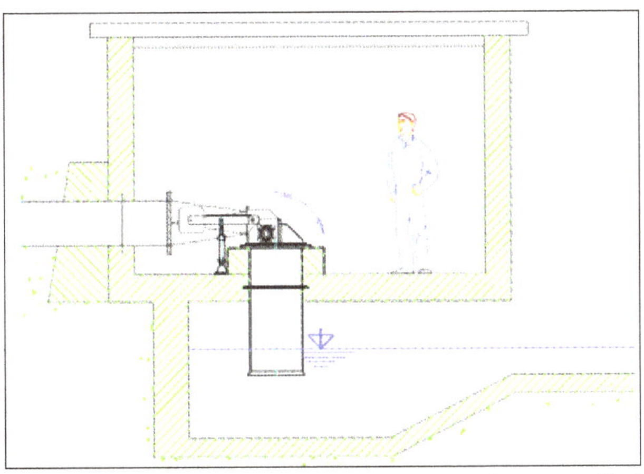

Cross-section of a typical low head crossflow turbine installation.

Crossflows are available in a number of rotor diameters, normally in 100 mm steps from 100 mm to 500 mm. The smaller diameters are for higher-head sites. For low head sites from 2.5 to 5 metres 300 mm diameter rotors are normally used. The 400 and 500 mm rotors are used on very low head sites. Renewables First have successfully installed 500 mm rotor crossflows on net heads as low as 1.75 metres. Many of the parts of crossflow turbines are standardised and only the width of the rotor is bespoke designed to match the expected range of flows at the hydro site.

Crossflow turbines require relatively little maintenance. The main rotor bearings are grease lubricated via two grease nipples, one on each side of the turbine and require greasing from a grease gun one per month. Annually the air inlet valve position should be checked. Hydro systems with crossflow turbines normally have 12 mm intake screens which prevent almost all but the tiniest debris entering the turbine, and this debris can pass through without any problem. Good quality crossflow turbines should operate efficiently for at least 40 years, and the oldest turbines are 60 years old and still operating reliably and efficiently and in most cases have only required new main bearings and guide vane bushes.

On the outside of the turbine the two inlet guide-vanes are moved via 'actuator arms' which are moved either by electric or hydraulic actuators. The amount and direction of actuator movement is governed by the system controller.

The turbine rotor normally rotates more slowly that the generator, so either a belt-drive or a gearbox is used to increase the speed. On smaller systems the drive pulley can be attached directly to the turbine shaft, and on larger systems an intermediate layshaft is used to avoid applying excessive radial loads onto the turbine drive shaft. The best (and most expensive) option is to use a gearbox which connects to the turbine via a flexible coupling, as shown on figure. The following pictures show a range of different crossflow turbines installed by Renewables First over the years.

Small Ecowave crossflow turbine.

Ossberger crossflow turbine.

Crossflow turbine with a gear-box drive system.

Axial Flow Turbine

The axial flow turbines have their axis of rotation oriented along the incoming water flow. It may be noted here that the bulb turbines explained above is also an axial flow turbine. These turbines have two or more radial blades. Construction of axial flow tidal stream turbines is similar to modern wind turbines seen today.

Artistic impression of an axial flow tidal stream turbine under development.

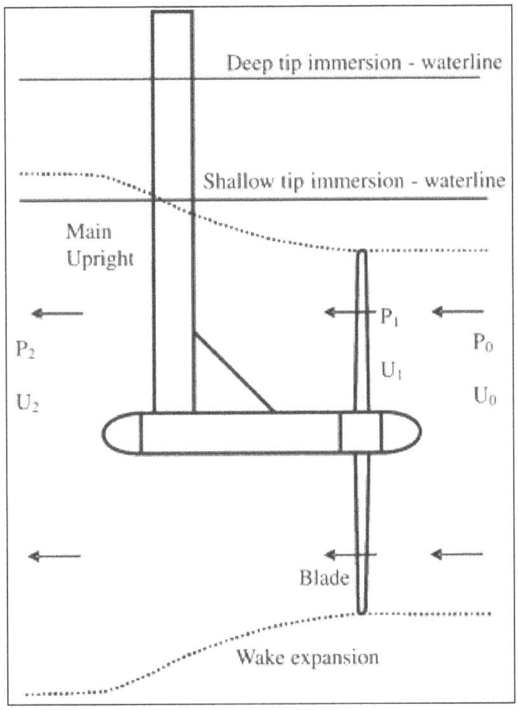

Axial flow tidal stream turbine.

Axial flow turbines have be oriented continuously in order to maximize energy capture by aligning their rotational axis with the direction of water flow. However the direction of rotation of turbines of this class changes as the tidal stream reverses the flow of water with change in tide. Also all turbines of this type experience huge drag which has to be countered appropriately by thrust bearings. The mast holding these turbines has to be designed to withstand the overturning drag moment.

Dynamic Tidal Power

Dynamic tidal power is a technology that uses the difference between the potential energy and kinetic energy of tides. Long dams are built from coasts straight out into the sea or ocean—meaning that the tides in the areas where these systems might be implemented usually flow parallel to their respective coasts. During the movement of tides, water on one side of the dam is at a higher level than the other side. As this water flows through the dam it drives a series of turbines installed within the dam and generates electricity. Furthermore, these dams are designed with bi-directional turbines, which flip 1800 after each tide in order to generate power both when the tide comes in and goes out. The added output from having bi-directional turbines is a huge advantage for these types of systems since they allow the power output to basically double.

Dynamic tidal power generation remains in the developmental stages meaning that any data is based on models and known tidal behaviour. With that being said, these models tend to be highly advanced and have high confidence in these theoretical predictions.

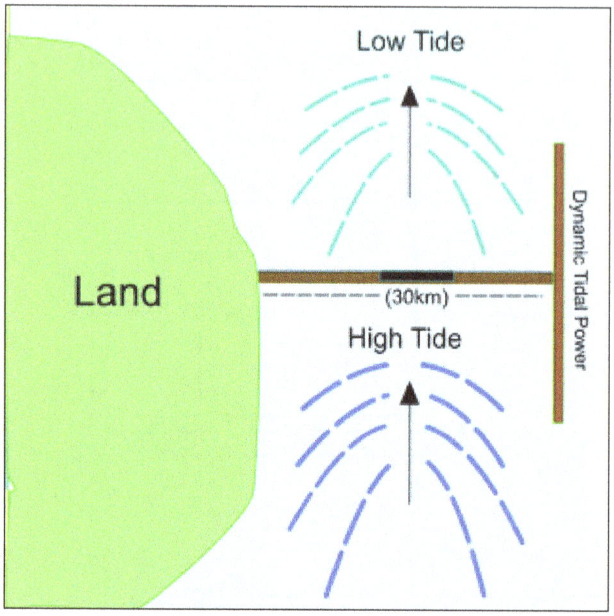

A representation of how Dynamic tidal power is used.

A dynamic tidal power dam can be 30 to 60 km long and is typically built perpendicular to the coast, running relatively straight out into the ocean. This design does not enclose any area, thus allows more freedom for aquatic life without risk of being trapped. These dams generate a water level difference from one side of the dam to the other—known as 'head'. The head difference (difference in water level between both sides) is predicted to be able to reach up to a few meters. The higher the head difference will result in much higher power production.

Power Production

In 2015, Canada generated 670 terawatt-hours of electricity. This means that a single dynamic tidal power dam, with an 8 GW installed capacity (this is the theoretical maximum output of a power plant) and a capacity factor of 30% (this is the ratio of actual power output to theoretical power output) could generate almost 4% of Canada's total power output or 21 terawatt-hours annually.

Positives of Dynamic Tidal Power

Multiple Functions

These types of dams, by necessity, tend to be very long (30 to 60 km) which is significantly longer (and more expensive to build) than current coastal structures. However, a long dam can function with multiple purposes such as land reclamation, connections between islands, mainland coastal protection, and deep sea and liquefied natural gas ports. The initial investment costs could easily be shared between these different industries, thus helping to lower the cost per kWh and create higher return investments.

Stable, Sustainable and Available

The moon (and subsequent tidal force) isn't going anywhere anytime soon which means that this type of tidal power will be around indefinitely. Furthermore, the tidal cycle is predictable and repetitive which allows us to know exactly when and how much power is obtainable.

Interestingly enough, since these dams are built perpendicular to the shore and the tide flows parallel, dynamic tidal power doesn't require huge tidal ranges. Therefore, these could be implemented in locations throughout the world, which is advantageous over other forms of tidal power (ex. tidal stream generators and tidal barrages) since they tend to require more intense tidal fluctuations in order to be economically viable.

Negatives of Dynamic Tidal Power

Lack of Proof

Unfortunately, due to the nature of the tides and the design of these dams, a full project must be built in order to guarantee predicted success. Building a miniature dam would be ineffective since these dams require a predicted minimum length of 30 km to ensure economic viability. Such a dam would require an enormous capital investment and if the design failed a lot of money would be lost. Furthermore, construction of a structure that extends 30 km or further into the ocean has never been attempted—the technology is available but the difficulty is extremely high.

Environmental Impact

The environmental impact of such a system could be significant. Unfortunately, in areas where any tidal power stations might be built happen to be some of the most densely populated ecosystems in the ocean. Changes in water levels might harm plant and animal life and alter the sea water composition. Also, turbines in dynamic tidal power stations move quickly, and if protection is not built in, marine animals can be caught in the blades. If aquatic life drops in the area, birds typically found there might migrate to different places. An entire major ecosystem could be significantly altered. Further research is needed in order to determine just how large of an impact such stations might have.

High Costs

It is difficult to determine the cost per MW for these types of power stations since none have been built yet. However, the costs are sure to be much higher than any other field of energy generation.

While the costs are high and the results as yet unknown, this technology may still have a place in the future electrical grid. As technology advances (which it will) and as climate change becomes more prevalent (which is virtually certain) there will be additional pressure to look for power with lower greenhouse gas emissions. Thus, the more that is known about alternative energy resources, the lower their costs. With these lower costs, transitions in power generation will be smoother.

Ocean Thermal Energy Conversion

Ocean thermal energy conversion (OTEC) generates electricity indirectly from solar energy by harnessing the temperature difference between the sun-warmed surface of tropical oceans and the colder deep waters. A significant fraction of solar radiation incident on the ocean is retained by seawater in tropical regions, resulting in average year-round surface temperatures of about 28 °C. Deep, cold water, meanwhile, forms at higher latitudes and descends to Sow along the seafloor toward the equator. The warm surface layer, which extends to depths of about 100-200m, is separated from the deep cold water by a thermocline. The temperature difference, DT, between the surface and thousand-meter depth ranges from 10 to 25 °C, with larger differences occurring in equatorial and tropical waters, as depicted in figure. DT establishes the limits of the performance of OTEC power cycles; the rule-of thumb is that a differential of about 20 °C is necessary to sustain viable operation of an OTEC facility.

Since OTEC exploits renewable solar energy, recurring costs to generate electrical power are minimal. However, the fixed or capital costs of OTEC systems per kilowatt of generating capacity are very high because large pipelines and heat exchangers are needed to produce relatively modest amounts of electricity. These high fixed costs dominate the economics of OTEC to the extent that it currently cannot compete with conventional power systems, except in limited niche markets. Considerable effort has been expended over the past two decades to develop OTEC by-products, such as fresh water, air conditioning, and mariculture that could offset the cost penalty of electricity generation.

State of the Technology

OTEC power systems operate as cyclic heat engines. They receive thermal energy through heat transfer from surface sea water warmed by the sun, and transform a portion of this energy to electrical power. The Second Law of Thermodynamics precludes the complete conversion of thermal energy in to electricity. A portion of the heat extracted from the warm sea water must be rejected to a colder thermal sink. The thermal sink employed by OTEC systems is sea water drawn from the ocean depths by means of a submerged pipeline. A steady-state control volume energy analysis yields the result that net electrical power produced by the engine must equal the difference between the rates of heat transfer from the warm surface water and to the cold deep water. The limiting (i.e., maximum) theoretical Carnot energy conversion efficiency of a cyclic heat engine scales with the difference between the temperatures at which these heat transfers occur. For OTEC, this difference is determined by DT and is very small; hence, OTEC efficiency is low. Although viable OTEC systems are characterized by Carnot efficiencies in the range of 6-8%, state-of-the-art combustion steam power cycles, which tap much higher temperature energy sources, are theoretically capable of converting more than 60% of the extracted thermal energy into electricity.

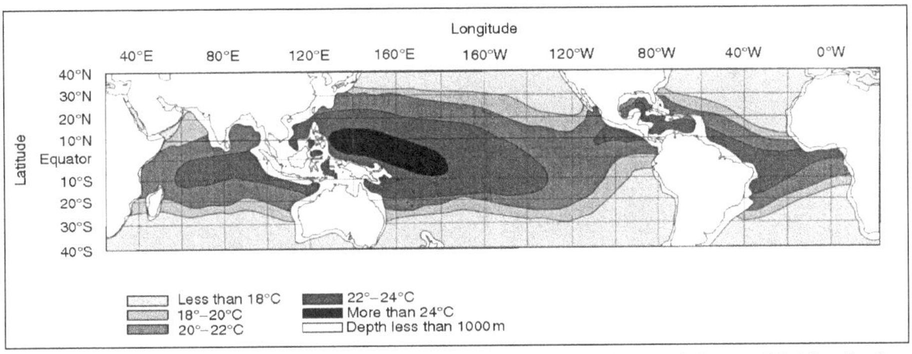

Temperature difference between surface and deep sea water in regions of the world. The darkest areas have the greatest temperature difference and are the best locations for OTEC systems.

The low energy conversion efficiency of OTEC means that more than 90% of the thermal energy extracted from the ocean's surface is 'wasted' and must be rejected to the cold, deep sea water. This necessitates large heat exchangers and seawater Sow rates to produce relatively small amounts of electricity.

In spite of its inherent inefficiency, OTEC, unlike conventional fossil energy systems, utilizes a renewable resource and poses minimal threat to the environment. In fact, it has been suggested that widespread adoption of OTEC could yield tangible environmental benefits through avenues such as reduction of greenhouse gas CO_2 emissions; enhanced uptake of atmospheric CO_2 by marine organism populations sustained by the nutrient-rich, deep OTEC sea water; and preservation of corals and hurricane amelioration by limiting temperature rise in the surface ocean through energy extraction and artificial upwelling of deep water.

Carnot efficiency applies only to an ideal heat engine. In real power generation systems, irreversibilities will further degrade performance. Given its low theoretical efficiency, successful implementation of OTEC power generation demands careful engineering to minimize irreversibilities. Although OTEC consumes what is essentially a free resource, poor thermodynamic performance will reduce the quantity of electricity available for sale and, hence, negatively affect the economic feasibility of an OTEC facility.

An OTEC heat engine may be configured following designs by J.A. D'Arsonval, the French engineer who first proposed the OTEC concept in 1881, or G. Claude, D'Arsonval's former student. Their designs are known, respectively, as closed cycle and open cycle OTEC.

Closed Cycle OTEC

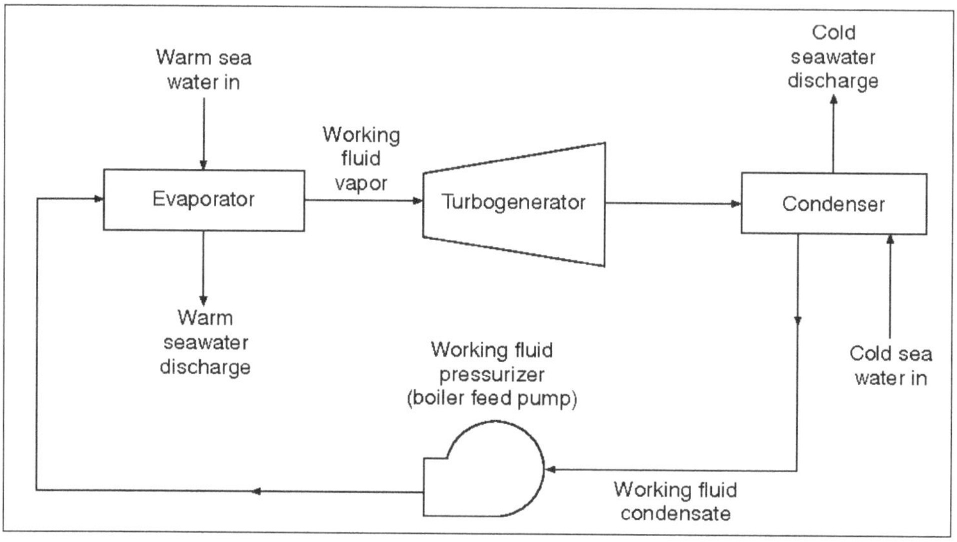

D'Arsonval's original concept employed a pure working fluid that would evaporate at the temperature of warm sea water. The vapor would subsequently expand and do work before being condensed by the cold sea water. This series of steps would be repeated continuously with the same working fluid, whose Sow path and thermodynamic process representation constituted closed loops - hence, the name 'closed cycle.' The specific process adopted for closed cycle OTEC is the Rankine, or vapor power, cycle. Figure is a simplified schematic diagram of a closed cycle OTEC system. The principal components are the heat exchangers, turbo generator, and seawater supply system, which, although not shown, accounts for most of the parasitic power consumption and a significant fraction of the capital expense. Also not included are ancillary devices such as separators to remove residual liquid downstream of the evaporator and subsystems to hold and supply working fluid lost through leaks or contamination.

In this system, heat transfer from warm surface sea water occurs in the evaporator, producing a saturated vapor from the working fluid. Electricity is generated when this gas expands to lower pressure through the turbine. Latent heat is transferred from the vapor to the cold sea water in the condenser and the resulting liquid is pressurized with a pump to repeat the cycle.

Schematic diagram of a closed-cycle OTEC system. The working fluid is vaporized by heat transfer from the warm sea water in the evaporator. The vapor expands through the turbogenerator and is condensed by heat transfer to cold sea water in the condenser. Closed-cycle OTEC power systems, which operate at elevated pressures, require smaller turbines than open-cycle systems.

The success of the Rankine cycle is a consequence of more energy being recovered when the vapor expands through the turbine than is consumed in re-pressurizing the liquid. In conventional (e.g., combustion) Rankine systems, this yields net electrical power. For OTEC, however, the remaining balance may be reduced substantially by an amount needed to pump large volumes of sea water through the heat exchangers. (One misconception about OTEC is that tremendous energy must be expended to bring cold sea water up from depths approaching 1000 meters. In reality, the natural hydrostatic pressure gradient provides for most of the increase in the gravitational potential energy of a fluid particle moving with the gradient from the ocean depths to the surface).

Irreversibilities in the turbomachinery and heat exchangers reduce cycle efficiency below the Carnot value. Irreversibilities in the heat exchangers occur when energy is transferred over a large temperature difference. It is important, therefore, to select a working fluid that will undergo the desired phase changes at temperatures established by the surface and deep sea water. Insofar as a large number of substances can meet this requirement (because pressures and the pressure ratio across the turbine and pump are design parameters), other factors must be considered in the selection of a working fluid including: cost and availability, compatibility with system materials, toxicity, and environmental hazard. Leading candidate working fluids for closed cycle OTEC applications are ammonia and various fluorocarbon refrigerants. Their primary disadvantage is the environmental hazard posed by leakage; ammonia is toxic in moderate concentrations and certain fluorocarbons have been banned by the Montreal Protocol because they deplete stratospheric ozone.

The Kalina, or adjustable proportion fluid mixture (APFM), cycle is a variant of the OTEC closed cycle. Whereas simple closed cycle OTEC systems use a pure working fluid, the Kalina cycle proposes to employ a mixture of ammonia and water with varying proportions at different points in the system. The advantage of a binary mixture is that, at a given pressure, evaporation or condensation occurs over a range of temperatures; a pure fluid, on the other hand, changes phase at constant temperature. This additional degree of freedom allows heat transfer-related irreversibilities in the evaporator and condenser to be reduced.

Although it improves efficiency, the Kalina cycle needs additional capital equipment and may impose severe demands on the evaporator and condenser. The efficiency improvement will require some combination of higher heat transfer coefficients, more heat transfer surface area, and increased seawater Sow rates. Each has an associated cost or power penalty. Additional analysis and testing are required to confirm whether the Kalina cycle and assorted variations are viable alternatives.

Open Cycle OTEC

Claude's concern about the cost and potential bio-fouling of closed cycle heat exchangers led him to propose using steam generated directly from the warm sea water as the OTEC working fluid. The steps of the Claude, or open, cycle are:

- Sash evaporation of warm sea water in a partial vacuum.
- Expansion of the steam through a turbine to generate power.
- Condensation of the vapor by direct contact heat transfer to cold sea water.
- Compression and discharge of the condensate and any residual noncondensable gases.

Unless fresh water is a desired by-product, open cycle OTEC eliminates the need for surface heat exchangers. The name 'open cycle' comes from the fact that the working fluid (steam) is discharged after a single pass and has different initial and final thermodynamic states; hence, the Sow path and process are 'open'.

The essential features of an open cycle OTEC system are presented in figure. The entire system, from evaporator to condenser, operates at partial vacuum, typically at pressures of 1-3% of atmospheric. Initial evacuation of the system and removal of noncondensable gases during operation are performed by the vacuum compressor, which, along with the sea water and discharge pumps, accounts for the bulk of the open cycle OTEC parasitic power consumption.

The low system pressures of open cycle OTEC are necessary to induce boiling of the warm sea water. Flash evaporation is accomplished by exposing the sea water to pressures below the saturation pressure corresponding to its temperature. This is usually accomplished by pumping it into an evacuated chamber through spouts designed to maximize heat and mass transfer surface area. Removal of gases dissolved in the sea water, which will come out of solution in the low-pressure evaporator and compromise operation, may be performed at an intermediate pressure prior to evaporation.

Vapor produced in the Sash evaporator is relatively pure steam. The heat of vaporization is extracted from the liquid phase, lowering its temperature and preventing any further boiling. Flash evaporation may be perceived, then, as a transfer of thermal energy from the bulk of the warm sea water of the small fraction of mass that is vaporized. Less than 0.5% of the mass of warm sea water entering the evaporator is converted into steam.

The pressure drop across the turbine is established by the cold seawater temperature. At 4 °C, steam condenses at 813 Pa. The turbine (or turbine diffuser) exit pressure cannot fall below this value. Hence, the maximum turbine pressure drop is only about 3000Pa, corresponding to about a 3:1 pressure ratio. This will be further reduced to account for other pressure drops along the steam path and differences in the temperatures of the steam and seawater streams needed to facilitate heat transfer in the evaporator and condenser.

Condensation of the low-pressure steam leaving the turbine may employ a direct contact condenser (DCC), in which cold sea water is sprayed over the vapor, or a conventional surface condenser that physically separates the coolant and the condensate. DCCs are inexpensive and have good heat transfer characteristics because they lack a solid thermal boundary between the warm and cool fluids. Surface condensers are expensive and more difficult to maintain than DCCs; however, they produce a marketable freshwater by-product.

Effluent from the condenser must be discharged to the environment. Liquids are pressurized to ambient levels at the point of release by means of a pump, or, if the elevation of the condenser is suitably high, can be compressed hydrostatically. Noncondensable gases, which include any residual water vapor, dissolved gases that have come out of solution and air that may have leaked into the system, are removed by the vacuum compressor.

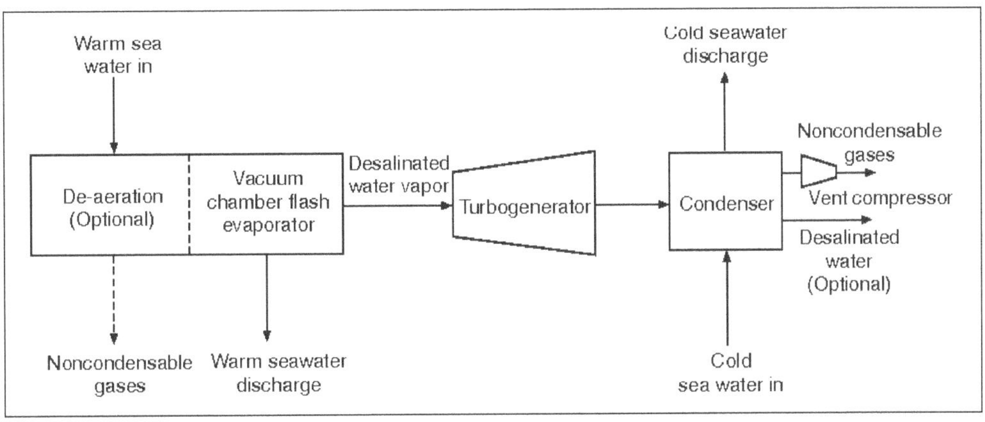

Figure shows schematic diagram of an open-cycle OTEC system. In open-cycle OTEC, warm sea water is used directly as the working fluid. Warm sea water is flash evaporated in a partial vacuum in the evaporator. The vapor expands through the turbine and is condensed with cold sea water. The principal disadvantage of open-cycle OTEC is the low system operating pressures, which necessitate large components to accommodate the high volumetric flow rates of steam.

Open cycle OTEC eliminates expensive heat exchangers at the cost of low system pressures. Partial vacuum operation has the disadvantage of making the system vulnerable to air in-leakage and promotes the evolution of noncondensable gases dissolved in sea water. Power must ultimately be expended to pressurize and remove these gases. Furthermore, as a consequence of the low steam density, volumetric Sow rates are very high per unit of electricity generated. Large components are needed to accommodate these Sow rates. In particular, only the largest conventional steam turbine stages have the potential for integration into open cycle OTEC systems of a few megawatts gross generating capacity. It is generally acknowledged that higher capacity plants will require a major turbine development effort.

The mist lift and foam lift OTEC systems are variants of the OTEC open cycle. Both employ the sea water directly to produce power. Unlike Claude's open cycle, lift cycles generate electricity with a hydraulic turbine. The energy expended by the liquid to drive the turbine is recovered from the warm sea water. In the lift process, warm seawater is Sash evaporated to produce a two-phase, liquid-vapor mixture - either a mist consisting of liquid droplets suspended in a vapor, or foam, where vapor bubbles are contained in a continuous liquid phase. The mixture rises, doing work against gravity. Here, the thermal energy of the vapor is expended to increase the potential energy of the fluid. The vapor is then condensed with cold sea water and discharged back into the ocean. Flow of the liquid through the hydraulic turbine may occur before or after the lift process. Advocates of the mist and foam lift cycles contend that they are cheaper to implement than closed cycle OTEC because they require no expensive heat exchangers, and are superior to the Claude cycle because they utilize a hydraulic turbine rather than a low pressure steam turbine. These claims await verification.

Hybrid Cycle OTEC

Some marketing studies have suggested that OTEC systems that can provide both electricity and water may be able to penetrate the marketplace more readily than plants dedicated solely to power generation. Hybrid cycle OTEC was conceived as a response to these studies. Hybrid cycles combine the potable water production capabilities of open cycle OTEC with the potential for large electricity generation capacities offered by the closed cycle.

Several hybrid cycle variants have been proposed. Typically, as in the Claude cycle, warm surface seawater is Sash evaporated in a partial vacuum. This low pressure steam flows into a heat exchanger where it is employed to vaporize a pressurized, low-boiling-point fluid such as ammonia. During this process, most of the steam condenses, yielding desalinated potable water. The ammonia vapor Sows through a simple closed-cycle power loop and is condensed using cold sea water. The uncondensed steam and other gases exiting the ammonia evaporator may be further cooled by heat transfer to either the liquid ammonia leaving the ammonia condenser or cold sea water. The noncondensables are then compressed and discharged to the atmosphere.

Steam is used as an intermediary heat transfer medium between the warm sea water and the ammonia; consequently, the potential for biofouling in the ammonia evaporator is reduced significantly. Another advantage of the hybrid cycle related to freshwater production is that condensation occurs at significantly higher pressures than in an open cycle OTEC condenser, due to the elimination of the turbine from the steam Sow path. This may, in turn, yield some savings in the amount of power consumed to compress and discharge the noncondensable gases from the system. These savings (relative to a simple Claude cycle producing electricity and water), however, are offset by the additional back work of the closed-cycle ammonia pump.

One drawback of the hybrid cycle is that water production and power generation are closely coupled. Changes or problems in either the water or power subsystem will compromise performance of the other. Furthermore, there is a risk that the potable water may be contaminated by an ammonia leak. In response to these concerns, an alternative hybrid cycle has been proposed, comprising decoupled power and water production components. The basis for this concept lies in the fact that warm sea water leaving a closed cycle evaporator is still sufficiently warm, and cold seawater exiting the condenser is sufficiently cold, to sustain an independent freshwater production process.

The alternative hybrid cycle consists of a conventional closed-cycle OTEC system that produces electricity and a downstream Sash-evaporation-based desalination system. Water production and electricity generation can be adjusted independently, and either can operate should a subsystem fail or require servicing. The primary drawbacks are that the ammonia evaporator uses warm seawater directly and is subject to biofouling; and additional equipment, such as the potable water surface condenser, is required, thus increasing capital expenses.

Environmental Considerations

OTEC systems are, for the most part, environmentally benign. Although accidental leakage of closed cycle working fluids can pose a hazard, under normal conditions, the only effluents are the mixed seawater discharges and dissolved gases that come out of solution when sea water is depressurized. Although the quantities of outgassed species may be significant for large OTEC systems, with the exception of carbon dioxide, these species are benign. Carbon dioxide is a greenhouse gas and can impact global climate; however, OTEC systems release one or two orders of magnitude less carbon dioxide than comparable fossil fuel power plants and those emissions may be sequestered easily in the ocean or used to stimulate marine biomass production.

OTEC mixed seawater discharges will be at lower temperatures than sea water at the ocean surface. The discharges will also contain high concentrations of nutrients brought up with the deep sea water and may have a different salinity. It is important; therefore, that release back into the ocean is conducted in a manner that minimizes unintended changes to the ocean mixed layer biota and avoids inducing long-term surface temperature anomalies. Analyses of OTEC effluent plumes suggest that discharge at depths of 50-100m should be sufficient to ensure minimal impact on the ocean environment. Conversely, the nutrient-rich OTEC discharges could be exploited to sustain open-ocean mariculture.

Economics of OTEC

Studies conducted to date on the economic feasibility of OTEC systems suffer from the lack of reliable cost data. Commercialization of the technology is unlikely until a full-scale plant is constructed and operated continuously over an extended period to provide these data on capital and personnel and maintenance expenses.

Uncertainties in financial analyses notwithstanding, projections suggest very high first costs for OTEC power system components. Small land-based or near-shore floating plants in the 1-10 MW range, which would probably be constructed in rural island communities, may require expenditures of \$10 000-\$20000 (in 1995 US dollars) per kW of installed generating capacity. Although there appears to be favorable economies of scale, larger floating (closed cycle) plants in the 50-100 MW range are still anticipated to cost about \$5000kW^{-1}. This is well in excess of the \$1000-\$2000kW^{-1} of fossil fuel power stations.

To enhance the economics of OTEC power stations, various initiatives have been proposed based on marketable OTEC by- or co-products. OTEC proponents believe that the first commercial OTEC plants will be shore-based systems designed for use in developing Pacific island nations, where potable water is in short supply. Many of these sites would be receptive to opportunities for economic growth provided by OTEC-related industries.

Fresh Water

The condensate of the open and hybrid cycle OTEC systems is desalinated water, suitable for human consumption and agricultural uses. Analyses have suggested that first-generation OTEC plants, in the 1-10MW range, would serve the utility power needs of rural Pacific island communities, with the desalinated water by-product helping to offset the high cost of electricity produced by the system.

Refrigeration and Air Conditioning

The cold, deep sea water can be used to maintain cold storage spaces, and to provide air conditioning. The Natural Energy Laboratory of Hawaii Authority (NELHA), which manages the site of Hawaii's OTEC experiments, has air-conditioned its buildings by passing the cold sea water through heat exchangers. A new deep seawater utilization test facility in Okinawa also employs cold seawater air conditioning. Similar small-scale operations would be viable in other locales. Economic studies have been performed for larger metropolitan and resort applications. These studies indicate that air conditioning new developments, such as resort complexes, with cold seawater may be economically attractive even if utility-grid electricity is available.

Mariculture

The cold deep ocean waters are rich in nutrients and low in pathogens, and therefore provide an excellent medium for the cultivation of marine organisms. The 322-acre NELHA facility has been the base for successful mariculture research and development enterprises. The site has an array of cold water pipes, originally installed for the early OTEC research, but since used for mariculture. The cold water is applied to cultivate Sounder, opihi (limpet; a shellfish delicacy), oysters, lobsters, sea urchins, abalone, kelp, nori (a popular edible seaweed used in sushi), and macro- and microalgae. Although many of these ongoing endeavors are profitable, high-value products such as biopharmaceuticals, biopigments, and pearls will need to be advanced to realize the full potential of the deep water.

The cold sea water may have applications for open-ocean mariculture. Artificial upwelling of deep water has been suggested as a method of creating new fisheries and marine biomass plantations. Should development proceed, open-ocean cages can be eliminated and natural feeding would replace expensive feed, with temperature and nutrient differentials being used to keep the fish stock in the kept environment.

Agriculture

An idea initially proposed by University of Hawaii researchers involves the use of cold sea water for agriculture. This involves burying an array of cold water pipes in the ground near to the surface to create cool weather growing conditions not found in tropical environments. In addition to cooling the soil, the system also drip irrigates the crop via condensation of moisture in the air on the cold water pipes. Demonstrations have determined that strawberries and other spring crops and Sowers can be grown throughout the year in the tropics using this method.

Energy Carriers

Although the most common scenario is for OTEC energy to be converted into electricity and delivered directly to consumers, energy storage has been considered as an alternative, particularly in applications involving floating plants moored far offshore. Storage would also allow the export of OTEC energy to industrialized regions outside of the tropics. Long-term proposals have included the production of hydrogen gas via electrolysis, ammonia synthesis, and the development of shore-based mariculture systems or floating OTEC plant-ships as oceangoing farms. Such farms would cultivate marine biomass, for example, in the form of fast-growing kelp which could be converted thermo chemically into fuel and chemical co-products.

References

- Tidal-Stream-Turbine-Introduction-current-and-future-Tidal-power-stations-278850827: researchgate.net, Retrieved 13, August 2020
- Ocean-energy: manifestias.com, Retrieved 12, July 2020
- Crossflow-turbines, hydropower-learning-centre, hydropower: renewablesfirst.co.uk, Retrieved 17, June 2020
- Ency-Oceans, ency-6140: curry.eas.gatech.edu, Retrieved 09, January 2020
- Dynamic-tidal-power, encyclopedia: energyeducation.ca, Retrieved 30, March 2020

Permissions

All chapters in this book are published with permission under the Creative Commons Attribution Share Alike License or equivalent. Every chapter published in this book has been scrutinized by our experts. Their significance has been extensively debated. The topics covered herein carry significant information for a comprehensive understanding. They may even be implemented as practical applications or may be referred to as a beginning point for further studies.

We would like to thank the editorial team for lending their expertise to make the book truly unique. They have played a crucial role in the development of this book. Without their invaluable contributions this book wouldn't have been possible. They have made vital efforts to compile up to date information on the varied aspects of this subject to make this book a valuable addition to the collection of many professionals and students.

This book was conceptualized with the vision of imparting up-to-date and integrated information in this field. To ensure the same, a matchless editorial board was set up. Every individual on the board went through rigorous rounds of assessment to prove their worth. After which they invested a large part of their time researching and compiling the most relevant data for our readers.

The editorial board has been involved in producing this book since its inception. They have spent rigorous hours researching and exploring the diverse topics which have resulted in the successful publishing of this book. They have passed on their knowledge of decades through this book. To expedite this challenging task, the publisher supported the team at every step. A small team of assistant editors was also appointed to further simplify the editing procedure and attain best results for the readers.

Apart from the editorial board, the designing team has also invested a significant amount of their time in understanding the subject and creating the most relevant covers. They scrutinized every image to scout for the most suitable representation of the subject and create an appropriate cover for the book.

The publishing team has been an ardent support to the editorial, designing and production team. Their endless efforts to recruit the best for this project, has resulted in the accomplishment of this book. They are a veteran in the field of academics and their pool of knowledge is as vast as their experience in printing. Their expertise and guidance has proved useful at every step. Their uncompromising quality standards have made this book an exceptional effort. Their encouragement from time to time has been an inspiration for everyone.

The publisher and the editorial board hope that this book will prove to be a valuable piece of knowledge for students, practitioners and scholars across the globe.

Index

www.ingramcontent.com/pod-product-compliance
Lightning Source LLC
LaVergne TN
LVHW080613210926
848485LV00003B/150
9781639873197